D1135814

17th Edition

IET Wiring Regulations
Inspection, Testing and Certification

- Fully up-to-date with the latest amendments to the 17th Edition of the IET Wiring Regulations
- Simplifies the advice found in the Wiring Regulations, explaining how they apply to working practice for inspection, testing and certification
- Expert advice from an engineering training consultant, supported with colour diagrams, examples and key data

This popular guide clarifies the requirements for inspection and testing, explaining in clear language those parts of the Wiring Regulations that most need simplifying. In addition to the descriptive and diagrammatic test methods that are required, explanations of the theory and reasoning behind test procedures are given, together with useful tables for the comparison of test results. The book also provides essential information on the completion of electrical installation certificates, with a step-by-step guide on the entries that need to be made and where to source data.

With the coverage carefully matched to the syllabus of the City & Guilds Certificates in Inspection, Testing and Certification of Electrical Installations (2394-10 and 2395-10) and Fundamental Inspection, Testing and Initial Verification (2392-10), and containing sample papers and suggested solutions, it is also an ideal revision guide.

Brian Scaddan, I Eng, MIET, is a consultant for and an Honorary Member of City & Guilds with over 40 years' experience in Further Education and training. He is Director of Brian Scaddan Associates Ltd, an approved City & Guilds training centre offering courses on all aspects of electrical installation contracting including the C&G 2382-15, 2392-10, 2377-22, 2394-01, 2395-01 and 2396-01. He is also a leading author of books for these courses.

17th Edition
IET Wiring Regulations
Inspection, Testing and Certification

Eighth Edition

Brian Scaddan

Routledge
Taylor & Francis Group

LONDON AND NEW YORK

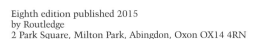

Eighth edition published 2015
by Routledge
2 Park Square, Milton Park, Abingdon, Oxon OX14 4RN

and by Routledge
711 Third Avenue, New York, NY 10017

Routledge is an imprint of the Taylor & Francis Group, an informa business

First edition published 1996 by Newnes, an imprint of Elsevier
Seventh edition published 2011 by Newnes, an imprint of Elsevier

British Library Cataloguing in Publication Data
A catalogue record for this book is available from the British Library

Library of Congress Cataloging-in-Publication Data
Scaddan, Brian.
17th edition IET wiring regulations. Inspection, testing and certification / Brian Scaddan. – 8th edition.
pages cm
Includes index.
1. Electric wiring, Interior–Safety regulations–Great Britain–Handbooks, manuals, etc. 2. Electric wiring, Interior–Insurance requirements–Great Britain–Handbooks, manuals, etc. 3. Electric wiring, Interior–Inspection–Handbooks, manuals, etc. 4. Electric wiring, Interior–Testing–Handbooks, manuals, etc. I. Title. II. Title: Inspection, testing, and certification. III. Title: IET wiring regulations, inspection, testing and certification.
TK3271.S2692 2015
621.319'24021841–dc23
2014048613

ISBN: 978-1-138-84886-3 (pbk)
ISBN: 978-1-315-72595-6 (ebk)

Typeset in Kuenstler 480 and Trade Gothic by
Servis Filmsetting Ltd, Stockport, Cheshire

Printed by Bell and Bain Ltd, Glasgow

Contents

Preface

This book is primarily for use as a study resource for the City & Guilds 2394 Initial Verification and C&G 2395 Periodic Inspection and Testing courses.

However, it is also a useful reference for C&G 2382-15, 17th Edition; C&G 2392-10, Fundamental Inspection and Testing, and C&G 2396, Design.

Brian Scaddan

Introduction

Many candidates embarking on the 2394 and 2395 courses find difficulty in either understanding what the Examiner is asking for in questions, or how to express themselves when answering. The sample papers and answers at the end of this book should help to alleviate these difficulties.

One of the common problems is a lack of understanding of some basic principles and incorrect use of technical wording. In many instances it may be unfair to quote 'misunderstanding' when it is probably just a case of having forgotten the details of the 17th Edition, etc.

Below is a list of common words, phrases and acceptable abbreviations that candidates really should be aware of, as they often appear in examination questions:

Statutory Documents
- The Health and Safety at Work Act (H&SWA). *Affects everyone at work.*
- The Electricity at Work Regulations (EAWR). *Affects those at work involved with electrical systems.*
- The Building Regulations Part 'P'. *Affects those who install electrical systems in domestic premises.*
- The Electricity Safety, Quality and Continuity Regulations (ESQCR). *These really only affect the suppliers of electrical energy to premises.*

Non-Statutory Documents
- BS 7671.
- The IET Guidance Note 3 (GN3). *This is specifically for Inspection and Testing.*
- The H&S Guidance Note GS38 (GS38). *This deals with electrical instruments, etc.*
- Any other documents that relate to inspection and testing.

Electrical System This is defined in the EAWR as anything that generates, stores, transmits, uses, etc. electrical energy (e.g. a power station or a torch battery or a test instrument, etc.).

Duty Holder This is the EAWR **Title** of anyone who has control of an electrical system.

Competent Person This is the EAWR **Status** of a Duty Holder.

Basic Protection Protection against electric shock under fault-free conditions (*touching an intentionally live part*)

Methods of Providing Basic Protection only:
- Insulation of live parts.
- Barriers or enclosures.
- Obstacles *(not common, only for use under the supervision of skilled persons)*.
- Placing out of arms' reach *(not common, only for use under the supervision of skilled persons)*.

Fault Protection Protection against electric shock under single-fault conditions *(touching a conductive part made live due to a fault)*.

Methods of Providing Fault Protection only:
- Automatic Disconnection of Supply (ADS). This is Earthing, Bonding and ensuring protective devices operate in the designated time.

Methods of Providing both Basic and Fault Protection:
- Double or Reinforced insulation.
- SELV or PELV.
- Electrical Separation (for one item of equipment, e.g. shaver point).

Exposed Conductive Part Casing of Class I equipment or metal conduit/trunking, etc.

Extraneous Conductive Part Structural steelwork, metallic gas, water, oil pipes, etc.

Additional Protection Used in the event of failure of Basic and/or Fault protection or carelessness by users.

Methods of Providing Additional Protection:
- RCD, 30 mA or less and operating within 40 ms at five times its rating
- Supplementary Equipotential Bonding.

Protective Conductors:
- **The Earthing conductor**
  Connects the Main Earthing Terminal (MET) to the means of earthing. *(The **Main** Earthing conductor is incorrect terminology.)*
- **Main Protective Bonding conductors**
  Connect the MET to extraneous conductive parts.
- **Circuit Protective conductors (cpc)**
  Connect the MET to exposed conductive parts.
- **Supplementary Protective Bonding conductors**
  Connect together exposed and extraneous conductive parts in locations such as bathrooms, swimming pools, etc. or where disconnection times cannot be met although RCDs are usually used in this case.

THE IET WIRING REGULATIONS BS 7671

Before we embark on the subject of inspection and testing, it is, perhaps, wise to examine in more detail some of the key topics previously listed.

Clearly, the protection of persons and livestock from shock and burns, etc. and the prevention of damage to property are priorities. In consequence, therefore, thorough inspection and testing of an installation and subsequent remedial work where necessary will significantly reduce the risks.

So let us start with electric shock; that is, the passage of current through the body of such magnitude as to have significant harmful effects. Figure 0.1 illustrates the generally accepted effects of current passing through the human body.

How then are we at risk of electric shock, and how do we protect against it?

There are two ways in which we can be at risk:

1. Touching live parts of equipment or systems that are intended to be live.
2. Touching conductive parts which are not meant to be live, but have become live due to a fault.

FIGURE 0.1   Shock levels.

1–2 mA	Barely perceptible, no harmful effects
5–10 mA	Throw off, painful sensation
10–15 mA	Muscular contraction, can't let go
20–30 mA	Impaired breathing
50 mA and above	Ventricular fibrillation and death

The conductive parts associated with the second of these can either be metalwork of electrical equipment and accessories (Class I) and that of electrical wiring systems such as metallic conduit and trunking, etc. called *exposed conductive parts*, or other metalwork such as pipes, radiators, girders, etc. called *extraneous conductive parts*.

Let us now consider how we may protect against electric shock from whatever source.

PROTECTION AGAINST SHOCK FROM BOTH TYPES OF CONTACT

One method of achieving this is by ensuring that the system voltage does not exceed extra low (50 V a.c., 120 V ripple-free d.c.), and that all associated wiring, etc. is separated from all other circuits of a higher voltage and earth. Such a system is known as a *separated extra low voltage (SELV)*. If a SELV system exceeds 25 V a.c., 60 V ripple-free d.c., then extra protection must be provided by barriers, enclosures and insulation.

BASIC PROTECTION

Apart from SELV, how can we prevent danger to persons and livestock from contact with intentionally live parts? Clearly we must minimize the risk of such contact, and this may be achieved in one or more of the following ways:

1. Insulate any live parts.
2. Ensure that any uninsulated live parts are housed in suitable enclosures and/or are behind barriers.
3. Place obstacles in the way. (This method would only be used in areas where skilled and/or authorized persons were involved.)
4. Placing live parts out of reach. (Once again, only used in special circumstances, e.g. live rails of overhead travelling cranes.)

A residual current device (RCD) may be used as additional protection to any of the other measures taken, provided that it is rated at 30 mA or less and has an operating time of not more than 40 ms at a test current of five times its operating current.

It should be noted that RCDs are not the panacea for all electrical ills, they can malfunction, but they are a valid and effective back-up to the other methods. They must not be used as the sole means of protection.

FAULT PROTECTION

How can we protect against shock from contact with unintentionally live, exposed or extraneous conductive parts whilst touching earth, or from contact between unintentionally live exposed and/or extraneous conductive parts? The most common method is by *protective earthing, protective equipotential bonding and automatic disconnection in case of a fault.*

All extraneous conductive parts are connected with a main protective bonding conductor and connected to the main earthing terminal, and all exposed conductive parts are connected to the main earthing terminal by the circuit protective conductors (cpc). Add to this overcurrent protection that will operate fast enough when a fault occurs and the risk of severe electric shock is significantly reduced.

Other means of fault protection may be used, but are less common and some require very strict supervision.

USE OF CLASS II EQUIPMENT

Often referred to as double-insulated equipment, this is typical of modern appliances where there is no provision for the connection of a cpc. This does not mean that there should be no exposed conductive parts and that the casing of equipment should be of an insulating material; it simply indicates that live parts are so well insulated that faults from live to conductive parts cannot occur.

NON-CONDUCTING LOCATION

This is basically an area in which the floor, walls and ceiling are all insulated. Within such an area there must be no protective conductors, and socket outlets will have no earthing connections.

It must not be possible simultaneously to touch two exposed conductive parts, or an exposed conductive part and an extraneous conductive part. This requirement clearly prevents shock current from passing through a person in the event of an earth fault, and the insulated construction prevents shock current from passing to earth.

EARTH-FREE LOCAL EQUIPOTENTIAL BONDING

This is in essence a Faraday cage, where all metals are bonded together but *not* to earth. Obviously, great care must be taken when entering such a zone in order to avoid differences in potential between inside and outside.

The areas mentioned in this and the previous method are very uncommon. Where they do exist, they should be under constant supervision to ensure that no additions or alterations can lessen the protection intended.

ELECTRICAL SEPARATION

This method relies on a supply from a safety source such as an isolating transformer to BS EN 61558-2-6 which has no earth connection on the secondary side. In the event of a circuit that is supplied from a source developing a live fault to an exposed conductive part, there would be no path for shock current to flow (see Figure 0.2).

FIGURE 0.2 Electrical separation.

Table 0.1 IP Codes

First Numeral	Mechanical Protection
0	No protection of persons against contact with live or moving parts inside the enclosure. No protection of equipment against ingress of solid foreign bodies.
1	Protection against accidental or inadvertent contact with live or moving parts inside the enclosure by a large surface of the human body, for example, a hand, but not protection against deliberate access to such parts.
2	Protection against ingress of large solid foreign bodies. Protection against contact with live or moving parts inside the enclosure by fingers. Protection against ingress of medium-size solid foreign bodies.
3	Protection against contact with live or moving parts inside the enclosures by tools, wires or such objects of thickness greater than 2.5 mm. Protection against ingress of small foreign bodies.
4	Protection against contact with live or moving parts inside the enclosure by tools, wires or such objects of thickness greater than 1 mm. Protection against ingress of small-size solid foreign bodies.
5	Complete protection against contact with live or moving parts inside the enclosure. Protection against harmful deposits of dust. The ingress of dust is not totally prevented, but dust cannot enter in an amount sufficient to interfere with satisfactory operation of the equipment enclosed.
6	Complete protection against contact with live or moving parts inside the enclosures. Protection against ingress of dust.
Second Numeral	**Liquid Protection**
0	No protection.
1	Protection against drops of condensed water. Drops of condensed water falling on the enclosure shall have no harmful effect.
2	Protection against drops of liquid. Drops of falling liquid shall have no harmful effect when the enclosure is tilted at any angle up to 15° from the vertical.
3	Protection against rain. Water falling in rain at an angle equal to or smaller than 60° with respect to the vertical shall have no harmful effect.
4	Protection against splashing. Liquid splashed from any direction shall have no harmful effect.
5	Protection against water jets. Water projected by a nozzle from any direction under stated conditions shall have no harmful effect.

Second Numeral	Liquid Protection
6	Protection against conditions on ships' decks (deck with watertight equipment). Water from heavy seas shall not enter the enclosures under prescribed conditions.
7	Protection against immersion in water. It must not be possible for water to enter the enclosure under stated conditions of pressure and time.
8	Protection against indefinite immersion in water under specified pressure. It must not be possible for water to enter the enclosure.
X	Indicates no specified protection.

Once again, great care must be taken to maintain the integrity of this type of system, as an inadvertent connection to earth, or interconnection with other circuits, would render the protection useless.

Additional protection by RCDs is a useful back-up to other methods of shock protection.

The use of enclosures is not limited to protection against shock from contact with live parts; they clearly provide protection against the ingress of foreign bodies and moisture. In order to establish to what degree an enclosure can resist such ingress, reference to the Index of Protection (IP) code (BS EN 60529) should be made. Table 0.1 illustrates part of the IP code.

The most commonly quoted IP codes in the 17th edition are IPXXB or IP2X, and IPXXD or IP4X. The X denotes that protection is not specified, not that there is no protection. For example, an enclosure that was to be immersed in water would be classified IPX8, there would be no point using the code IP68.

Note

IPXXB denotes protection against finger contact only.
IPXXD denotes protection against penetration by 1 mm diameter wires only.

CHAPTER 1

An Overview

Important terms/topics covered in this chapter:

- Statutory and Non-Statutory Regulations
- Electrical systems
- The Building Regulations Part 'P'
- Instruments

By the end of this chapter the reader should:

- be aware of the Statutory and Non-Statutory Regulations that are relevant to installation work,
- know the range of instruments required,
- know the requirements regarding the use and performance of test equipment.

So, here you are outside the premises, armed with lots of test instruments, a clipboard, a pad of documents that require completing, the IET Regulations, Guidance Notes 3 and an instruction to carry out an inspection and test of the electrical installation therein. Dead easy, you've been told, piece of cake, just poke about a bit, 'Megger' the wiring, write the results down, sign the test certificate and you should be onto the next job within the hour!

Oh! If only it were that simple! What if lethal defects were missed by just 'poking about'? What if other tests should have been carried out which may have revealed serious problems? What if things go wrong after you have signed to say all is in accordance with the Regulations? What if you were not actually competent to carry out the inspection and test in the first place? What if … and so on, the list is endless. Inspection, testing and certification is a serious and, in many instances, a complex matter, so let us wind the clock back to the point at which you were about to enter the premises to carry out your tests, and consider the implications of carrying out an inspection and test of an installation.

What are the legal requirements in all of this? Where do you stand if things go wrong? What do you need to do to ensure compliance with the law?

17th Edition IET Wiring Regulations: Inspection, Testing and Certification. 978-1-138-84886-3.

It is probably best at this point to consider the types of Inspection and Test that need to be conducted and the certification required.

There are two types:

1. Initial Verification.
2. Periodic Inspection and Testing.

Initial Verification is required for new work and alterations and additions (covered in City & Guilds 2392-10 and the more advanced 2394-01).

Periodic Inspection and Testing is required for existing installations (this and Initial Verification are covered in City & Guilds 2395-01).

The certification required for (1) (above) is an Electrical Installation Condition Report (EICR).

The certification required for (2) (above) is a Periodic Inspection Report (PIR). This could be referred to as a Condition Report.

Both must be accompanied by a schedule of test results and a schedule of inspections.

In the case of an addition or simple alteration that does not involve the installation of a new circuit (e.g. a spur from a ring final circuit), tests must be conducted but the certification required is a Minor Electrical Installation Works Certificate (MEIWC).

These are all covered in greater detail in Chapter 16.

STATUTORY AND NON-STATUTORY REGULATIONS

The statutory regulations that apply to electrical work are:
- The Health and Safety at Work Etc. Act (HSWA)
- The Electricity at Work Regulations (EAWR)
- The Building Regulations Part 'P' (applicable to domestic installations).

Non-statutory regulations include such documents as BS 7671:2008 and associated guidance notes, Guidance Note GS 38 on test equipment, etc.

The IET Wiring Regulations (BS 7671:2008) and associated guidance notes are *not* statutory documents; they can, however, be used in a court of law to prove compliance with statutory requirements such as the Electricity at Work Regulations (EAWR) 1989, which cover all work activity associated with electrical systems. A list of other statutory regulations is given in Appendix 2 of the IET Regulations. However, it is the EAWR that are most closely associated with BS 7671, and as such it is worth giving some areas a closer look.

In the EAWR there are 33 Regulations in all, 12 of which deal with the special requirements of mines and quarries, one which deals with extension outside Great Britain, and three which deal with effectively exemptions. We are only concerned with the first 16 Regulations, and Regulation 29, the defence regulation, which we shall come back to later. Let us start then with a comment on the meaning of *electrical systems and equipment*.

ELECTRICAL SYSTEMS AND EQUIPMENT

According to the EAWR, electrical systems and equipment can encompass anything from power stations to torch or wrist-watch batteries. A battery may not create a shock risk, but may cause burns or injury as a result of attempting to destroy it by fire, whereby explosions may occur. A system can actually include the source of energy, so a test instrument with its own supply, for example, a continuity tester, is a system in itself, and a loop impedance tester, which requires an external supply source, becomes part of the system into which it is connected. From the preceding comments it will be obvious then that, in broad terms, if something is electrical, it is or is part of an electrical system. So, where does responsibility lie for any involvement with such a system?

The EAWR requires that every employer, employee and self-employed person be responsible for compliance with the Regulations with regard to matters within their control, and as such are known as *duty holders*. Where then do you stand as the person about to conduct an inspection and test of an installation? Most certainly, you are a duty holder in that you have control of the installation insofar as you will ultimately pass the installation as safe or make recommendations to ensure its safety. You also have control of the test instruments which, as already stated, are systems in themselves, and control of the installation whilst testing is being carried out.

Any breach of the Regulations may result in prosecution, and unlike the other laws, under the EAWR you are presumed guilty and have to establish your innocence by invoking the Defence Regulation 29. Perhaps some explanation is needed here. Each of the 16 Regulations has a status, in that it is either *absolute* or *reasonably practicable*.

Regulations that are *absolute* must be conformed to at all cost, whereas those that are *reasonably practicable* are conformed to provided that all reasonable steps have been taken to ensure safety. For the contravention of an *absolute* requirement, Regulation 29 is available as a defence in the event of criminal prosecution, provided the accused can demonstrate that they took all reasonable and diligent steps to prevent danger or injury.

No one wants to end up in court accused of negligence, and so we need to be sure that we know what we are doing when we are inspecting and testing.

THE BUILDING REGULATIONS PART 'P'

Part 'P' of the building regulation requires that installations in dwellings be designed, installed and inspected and tested to the requirements of the current edition of the IET Wiring Regulations, BS 7671.

Details of some such work need to be notified to the Local Authority Building Control (LABC). This is a legal requirement.

Notifiable work

- Any new circuit fed from the consumer unit
- Replacement of a consumer unit
- Any work, additions or alterations in a bath/shower room, sauna or swimming pool.

Non-notifiable work

Any other work that involves replacements, additions or alterations anywhere else including gardens.

Certification

All electrical work has to be inspected, tested and certified with the issue of either:

- An Electrical Installation Certificate (EIC) for all new circuits and consumer unit replacements
- A Minor Electrical Installation Works Certificate (MEIWC) for small additions and/or alterations that do not comprise a new circuit
- An Electrical Installation Condition Report (EICR) where the inspector is *not* the installer.

If an installer belongs to a 'Competent Persons' scheme (i.e. is a Domestic Installer with an approval body, say, the NICEIC, ELECSA, etc.), he/she can self-certify the work and issue the appropriate certificate to the person ordering the work. The approval body will notify the LABC.

If an installer is not registered on a 'Competent Persons' scheme, but is qualified to inspect and test, he/she should issue the appropriate certificate to the LABC. They will take the inspector's qualifications into account and decide if any further action needs to be taken.

An installer who is not qualified to inspect and test may appoint a 'Registered Competent Person' to carry out the inspection and test. In this case only an EICR can be issued to the person ordering the work.

Apart from the knowledge required competently to carry out the verification process, the person conducting the inspection and test must be in possession of test instruments appropriate to the duty required of them.

INSTRUMENTS

In order to fulfil the basic requirements for testing to BS 7671, the following instruments are needed:

1. A low-resistance ohmmeter (continuity tester).
2. An insulation resistance tester.
3. A loop impedance tester.
4. A residual current device (RCD) tester.
5. A prospective fault current (PFC) tester.

6. An approved test lamp or voltage indicator.
7. A proving unit.
8. An earth electrode resistance tester.

Many instrument manufacturers have developed dual or multi-function instruments; hence it is quite common to have continuity and insulation resistance in one unit, loop impedance and PFC in one unit, loop impedance, PFC and RCD tests in one unit, etc. However, regardless of the various combinations, let us take a closer look at the individual test instrument requirements.

Low-resistance ohmmeters/continuity testers

Bells, buzzers, simple multimeters, etc. will all indicate whether or not a circuit is continuous, but will not show the difference between the resistance of, say, a 10 m length of 10 mm² conductor and a 10 m length of 1 mm² conductor. I use this example as an illustration, as it is based on a real experience of testing the continuity of a 10 mm² main protective bonding conductor between gas and water services. The services, some 10 m apart, were at either ends of a domestic premises. The 10 mm² conductor, connected to both services, disappeared under the floor, and a measurement between both ends indicated a resistance higher than expected. Further investigation revealed that just under the floor at each end, the 10 mm² conductor had been terminated in a connector block and the join between the two, about 8 m, had been wired with a 1 mm² conductor. Only a milli-ohmmeter would have detected such a fault.

A low-resistance ohmmeter should have a no-load source voltage of between 4 V and 24 V a.c. or d.c., and be capable of delivering an a.c. or d.c. short-circuit current of not less than 200 mA. It should have a resolution (i.e. a detectable difference in resistance) of at least 0.01 mV.

Insulation resistance testers

An *insulation resistance test* is the correct term for this form of testing, not a *megger* test, as megger is a manufacturer's trade name, not the name of the test.

> An insulation resistance tester must be capable of delivering 1 mA when the required test voltage is applied across the minimum acceptable value of insulation resistance.

Hence, an instrument selected for use on a low-voltage (50 V a.c.–1000 V a.c.) system should be capable of delivering 1 mA at 500 V across a resistance of 1 MΩ.

Loop impedance tester

This instrument functions by creating, in effect, an earth fault for a brief moment, and is connected to the circuit via a plug or by 'flying leads' connected separately to line, neutral and earth.

> The instrument should only allow an earth fault to exist for a maximum of 40 ms, and a resolution of 0.01 V is adequate for circuits up to 50 A. Above this circuit rating, the ohmic values become too small to give such accuracy using a standard instrument, and more specialized equipment may be required.

RCD tester

Usually connected by the use of a plug, although 'flying leads' are needed for non-socket outlet circuits, this instrument allows a range of out-of-balance currents to flow through the RCD to cause its operation within specified time limits.

> The test instrument should not be operated for longer than 2 s, and it should have a 10 per cent accuracy across the full range of test currents.

Earth electrode resistance tester

This is a 3- or 4-terminal, battery-powered resistance tester. Its application is discussed in Chapter 4.

PFC tester

This is either part of a combined PFC/Loop tester or a multi-function instrument. It is used to measure Prospective Short-Circuit Current (PSCC) line to neutral, or Prospective Earth Fault Current (PEFC) line to earth.

Approved test lamp or voltage indicator

A flexible cord with a lamp attached is not an approved device, nor for that matter is the ubiquitous 'testascope' or 'neon screwdriver', which encourages the passage of current, at low voltage, through the body!

A typical approved test lamp is shown in Figure 1.1.

The Health and Safety Executive, Guidance Note GS 38, recommends that the leads and probes associated with test lamps, voltage indicators, voltmeters, etc. have the following characteristics:

1. The leads should be adequately insulated and, ideally, fused.
2. The leads should be easily distinguished from each other by colour.
3. The leads should be flexible and sufficiently long for their purpose.
4. The probes should incorporate finger barriers, to prevent accidental contact with live parts.
5. The probes should be insulated and have a maximum of 2 mm of exposed metal, but preferably have spring-loaded enclosed tips.

Proving unit

This is an optional item of test equipment, in that test lamps should be proved on a known live supply which could, of course, be an adjacent socket or lighting point, etc. However, to prove a test lamp on such a known live supply may involve entry into enclosures with the associated hazards that such entry could bring. A proving unit is a compact device not much larger than a cigarette packet, which is capable of electronically developing 230 V d.c. across which the test lamp may be proved. The exception to this are test lamps incorporating 230 V lamps which will not activate from the small power source of the proving unit.

FIGURE 1.1 Approved test lamp.

Test lamps must be proved against a voltage similar to that to be tested. Hence, proving test lamps that incorporate an internal check; that is, shorting out the probes to make a buzzer sound is not acceptable if the voltage to be tested is higher than that delivered by the test lamp.

Care of test instruments

The EAWR (1989) requires that all electrical systems, and this includes test instruments, be maintained to prevent danger. This does not restrict such maintenance to just a yearly calibration, but requires equipment to be kept in good condition in order that it is safe to use at all times. In consequence it is important to ensure the continual accuracy of instruments by comparing test readings against known values. This is most conveniently achieved by the use of 'checkboxes' which are readily available.

Whilst test instruments and associated leads, probes and clips, etc. used in the electrical contracting industry are robust in design and manufacture, they still need treating with care and protecting from mechanical damage. Keep test gear in a separate box or case away from tools and sharp objects, and always check the general condition of a tester and leads before they are used.

Questions

1. State the statutory document most relevant to inspection and testing.
2. What is the minimum short-circuit current to be delivered by a low resistance ohmmeter?
3. What current must be delivered by an insulation resistance tester when used at 500 V across a resistance of 1 MΩ?
4. State the two tests carried out by a PFC tester.
5. What is the maximum length of exposed tip on the leads of a voltage indicator?

Answers

1. Electricity at Work Regulations (1989).
2. 200 mA.
3. 1 mA.
4. Prospective Short-Circuit Current (PSCC) and Prospective Earth Fault Current (PEFC).
5. 2 mm.

CHAPTER 2

Initial Verification

Important terms/topics covered in this chapter:

- Initial verification documentation
- Sequence of tests
- Inspection checklist

By the end of this chapter the reader should:

- know the correct sequence of tests to be carried out,
- be aware of the information required by an inspector,
- be aware of the extent of the inspections required.

CIRCUMSTANCES WHICH REQUIRE AN INITIAL VERIFICATION

New installations or additions or alterations.

GENERAL REASONS FOR INITIAL VERIFICATION

1. To ensure equipment and accessories are to a relevant standard.
2. To prove compliance with BS 7671.
3. To ensure that the installation is not damaged so as to impair safety.

INFORMATION REQUIRED

Assessment of general characteristics sections 311, 312 and 313 together with information such as drawings, charts, etc., in accordance with Regulation 514.9.1 (see BS 7671:2008).

DOCUMENTATION REQUIRED AND TO BE COMPLETED

Electrical Installation Certificate (EIC) signed or authenticated for the design and construction and then for the inspection and test (could be the same person). A schedule of test results and a schedule of inspections must accompany an EIC.

17th Edition IET Wiring Regulations: Inspection, Testing and Certification. 978-1-138-84886-3.

SEQUENCE OF TESTS

The IET Regulations indicate a preferred sequence of tests and state that if, due to a defect, compliance cannot be achieved, the defect should be rectified and the test sequence started from the beginning. The tests for 'Protection by separation' and 'Insulation of non-conducting floors and walls' all require specialist equipment and in consequence will not be discussed here. The sequence of tests for an initial inspection and test is as follows:

1. Continuity of protective conductors.
2. Continuity of ring final circuit conductors.
3. Insulation resistance.
4. Protection against direct contact by barriers or enclosures.
5. Polarity.
6. Earth electrode resistance.
7. Earth fault loop impedance.
8. Additional protection (RCDs).
9. Prospective fault current between live conductors and to earth.
10. Phase sequence.
11. Functional testing.
12. Voltage drop (not normally required for initial verification).

BS 7671:2008 requires tests 1–5 to be carried out in that order before the installation is energized and, if there is an earth electrode, its testing should be included. It does not require the live tests 7–11 to follow a sequence and item 12 is not usually required for an initial verification.

Even though no sequence is specified, it would always be appropriate to conduct test 7 before test 8 as high values of loop impedance or the absence of an earth path could result in dangerous voltages appearing between exposed and extraneous conductive parts and persons should be told not to touch metalwork whilst the test is being conducted.

One other test not included in Part 6 of the IET Regulations but which nevertheless has to be carried out is external earth fault loop impedance (Z_e).

Before any testing is carried out, a detailed physical inspection must be made to ensure that all equipment is to a relevant British or Harmonized

European Standard, that it is erected/installed in compliance with the IET Regulations, and that it is not damaged such that it could cause danger. In order to comply with these requirements, the Regulations give a checklist of items that, where relevant, should be inspected.

However, before such an inspection, and test for that matter, is carried out, certain information *must* be available to the verifier. This information is the result of the assessment of fundamental principles BS 7671 Section 131 and the Assessment of General Characteristics required by IET Regulations Part 3, sections 311, 312, 313, and drawings, charts and similar information relating to the installation. It is at this point that most readers who work in the real world of electrical installation will be lying on the floor laughing hysterically.

Let us assume that the designer and installer of the installation are competent professionals, and all of the required documentation is available.

Interestingly, one of the items on the checklist *is* the presence of diagrams, instructions and similar information. If these are missing then there is a deviation from the Regulations.

Another item on the list is the verification of conductors for current-carrying capacity and voltage drop in accordance with the design. How on earth can this be verified without all the information? A 30 A Type B circuit breaker (CB) or Type 2 miniature circuit breaker (MCB) protecting a length of 4 mm^2 conductor may look reasonable, but is it correct, and are you prepared to sign to say that it is unless you are sure? Let us look then at some items that would need inspecting.

1. **Connection of conductors:** Are terminations electrically and mechanically sound? Is insulation and sheathing removed only to a minimum to allow satisfactory termination?
2. **Identification of conductors:** Are conductors correctly identified in accordance with the Regulations?
3. **Routing of cables:** Are cables installed such that account is taken of external influences such as mechanical damage, corrosion, heat, etc.?
4. **Conductor selection:** Are conductors selected for current-carrying capacity and voltage drop in accordance with the design?

5. **Connection of single pole devices:** Are single pole protective and switching devices connected in the line conductor only?

6. **Accessories and equipment:** Are all accessories and items of equipment correctly connected?

7. **Thermal effects:** Are fire barriers present where required and protection against thermal effects provided?

8. **Protection against shock:** What methods have been used to attain both basic protection and fault protection?

9. **Mutual detrimental influence:** Are wiring systems installed such that they can have no harmful effect on non-electrical systems, or those systems of different currents or voltages are segregated where necessary?

10. **Isolation and switching:** Are there appropriate devices for isolation and switching correctly located and installed?

11. **Undervoltage:** Where undervoltage may give rise for concern, are there protective devices present?

12. **Labelling:** Are all protective devices, switches (where necessary) and terminals correctly labelled?

13. **External influences:** Have all items of equipment and protective measures been selected in accordance with the appropriate external influences?

14. **Access:** Are all means of access to switchgear and equipment adequate?

15. **Notices and signs:** Are danger notices and warning signs present?

16. **Diagrams:** Are diagrams, instructions and similar information relating to the installation available?

17. **Erection methods:** Have all wiring systems, accessories and equipment been selected and installed in accordance with the requirements of the Regulations, and are fixings for equipment adequate for the environment?

All defects and omissions, etc. in new work must be rectified and, in the case of an addition such as, say, a shower circuit, defects found in the existing installation that are unrelated to the new work should be recorded on the EIC.

Questions

1. An installation is to have the following tests conducted: (1) loop impedance, (2) polarity, (3) ring circuit continuity, and (4) insulation resistance. What is the correct sequence for carrying out the tests?
2. Which test is not normally required for an initial verification?
3. The details of which sections of BS 7671 are required to be made available to a person carrying out inspection and testing of an installation?
4. What inspection checklist item relates to damage to cables?

Answers

1. (3), (4), (2), (1).
2. Voltage drop.
3. 131, 311, 312, 313.
4. Routing of cables.

CHAPTER 3

Testing Continuity of Protective Conductors (Low-Resistance Ohmmeter)

Important terms/topics covered by this chapter:

- Protective bonding conductors
- Circuit protective conductors
- Parallel earth paths
- $(R_1 + R_2)$ values

By the end of this chapter the reader should:

- know what test instrument to use,
- understand the importance of disconnecting protective conductors for testing,
- know the importance of isolation, where protective conductors cannot be disconnected,
- be aware of the effects of parallel earth paths,
- be able to determine the approximate value of a protective conductor, given its length,
- know the preferred method of cpc continuity testing,
- know why $(R_1 + R_2)$ values are important.

All protective conductors, including main protective and supplementary bonding conductors, must be tested for continuity using a low-resistance ohmmeter. A visual inspection is sufficient for short lengths where the conductors are visible throughout their length.

For main protective bonding conductors there is no single fixed value of resistance above which the conductor would be deemed unsuitable. Each measured value, if indeed it is measurable for very short lengths, should be compared with the relevant value for a particular conductor length and size. Such values are shown in Table 3.1.

Where a supplementary bonding conductor has been installed between *simultaneously accessible* exposed and extraneous conductive parts as an

Table 3.1	Resistance (in Ω) of Copper Conductors at 20°C								
CSA	**Length (m)**								
(mm²)	**5**	**10**	**15**	**20**	**25**	**30**	**35**	**40**	**45**	**50**
1.0	0.09	0.18	0.27	0.36	0.45	0.54	0.63	0.72	0.82	0.9
1.5	0.06	0.12	0.18	0.24	0.3	0.36	0.43	0.48	0.55	0.6
2.5	0.04	0.07	0.11	0.15	0.19	0.22	0.26	0.3	0.33	0.37
4.0	0.023	0.05	0.07	0.09	0.12	0.14	0.16	0.18	0.21	0.23
6.0	0.02	0.03	0.05	0.06	0.08	0.09	0.11	0.13	0.14	0.16
10.0	0.01	0.02	0.03	0.04	0.05	0.06	0.063	0.07	0.08	0.09
16.0	0.006	0.01	0.02	0.023	0.03	0.034	0.04	0.05	0.05	0.06
25.0	0.004	0.007	0.01	0.015	0.02	0.022	0.026	0.03	0.033	0.04
35.0	0.003	0.005	0.008	0.01	0.013	0.016	0.019	0.02	0.024	0.03

addition to fault protection and there is doubt as to the effectiveness of the equipotential bonding, then the resistance (R) of the conductor must be equal to or less than $50/I_a$. So, $R \leq 50/I_a$ where 50 is the voltage above which exposed metalwork should not rise, and I_a is the minimum current causing operation of the circuit protective device within 5 s.

For example, suppose a 45 A BS 3036 fuse protects a cooker circuit, the disconnection time for the circuit cannot be met, and so a supplementary bonding conductor has been installed between the cooker case and an adjacent central heating radiator. The resistance (R) of that conductor should not be greater than $50/I_a$, and I_a in this case is 145 A (see Figure 3.2B of the IET Regulations); that is, $50/145 = 0.34 \ \Omega$.

How then do we conduct a test to establish continuity of main or supplementary bonding conductors? Quite simple really: just connect the leads from a low-resistance ohmmeter to the ends of the bonding conductor (Figure 3.1). One end should be disconnected from its bonding clamp, otherwise any measurement may include the resistance of parallel paths of other earthed metalwork. Remember to zero/null the instrument first or, if this facility is not available, record the resistance of the test leads so that this value can be subtracted from the test reading.

FIGURE 3.1 Testing main protective bonding.

Important Note

If the installation is in operation, then *never* disconnect protective bonding conductors unless the supply can be isolated. Without isolation, persons and livestock are at risk of electric shock. In this instance, or where the connections to extraneous conductive parts are not accessible, the test is conducted either between the connected bonding conductors or between extraneous conductive parts. The resistance value obtained should be no greater than 0.05 Ω.

The continuity of circuit protective conductors may be established in the same way, but a second method is preferred, as the results of this second test indicate the value of $(R_1 + R_2)$ for the circuit in question.

The test is conducted in the following manner:

1. Temporarily link together the line conductor and cpc of the circuit concerned in the distribution board or consumer unit.
2. Test between line and cpc at each outlet in the circuit. A reading indicates continuity.
3. Record the test result obtained at the furthest point in the circuit. This value is $(R_1 + R_2)$ for the circuit, and is important for use with the formula $Z_s = Z_e + (R_1 + R_2)$ for confirming measured values

of Z_s or for calculation where Z_s cannot be measured. It should also be noted that for lighting circuits the test should be carried out at the switches, as these are the furthest point for each luminaire.

Figure 3.2 illustrates the above method.

There may be some difficulty in determining the $(R_1 + R_2)$ values of circuits in installations that comprise steel conduit and trunking, and/or steel-wire-armoured (SWA) and mineral-insulated metal-sheathed (MIMS) cables, because of the parallel earth paths that are likely to exist. In these cases, continuity tests may have to be carried out at the installation stage before accessories are connected or terminations made off as well as after completion.

Although it is no longer considered good working practice to use steel conduit or trunking as a protective conductor, it is permitted, and hence its continuity must be proved. The enclosure must be inspected along its length to ensure that it is sound and then the standard low-resistance test is performed.

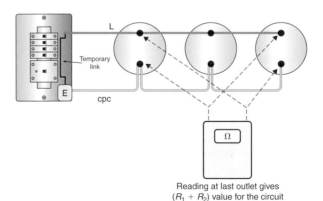

Reading at last outlet gives
$(R_1 + R_2)$ value for the circuit

FIGURE 3.2 Testing cpc continuity.

Questions

1. What instrument is used for testing the continuity of protective conductors?
2. What would be the approximate resistance value of a 10 mm^2 protective bonding conductor, 15 m long?
3. What may be the effect on a resistance test reading taken between the connected ends of a protective bonding conductor?
4. Where, on a lighting circuit, should a cpc continuity test be conducted?
5. What is the significance of the reading at the end of the circuit in Q.4 above?
6. Why is a value of $(R_1 + R_2)$ important, other than confirming cpc continuity?

Answers

1. Low-resistance ohmmeter.
2. $0.03\ \Omega$.
3. A lower value of resistance than the actual conductor value due to parallel earth paths.
4. At all points on the circuit.
5. It is $(R_1 + R_2)$ for the circuit.
6. It can be used in the formula $Z_s = Z_e + (R_1 + R_2)$ to confirm a measured value of Z_s or to calculate a Z_s value where measurement is not possible.

CHAPTER 4

Testing Continuity of Ring Final Circuit Conductors (Low-Resistance Ohmmeter)

Important terms/topics covered in this chapter:

- Low-resistance ohmmeter
- Ring final circuit interconnections
- Spurs
- $(R_1 + R_2)$ values
- Interpretation of test values

By the end of this chapter the reader should:

- know the reasons for conducting a ring final circuit continuity test,
- understand the problems that interconnections may create,
- understand why initial conductor cross-connections are made for the test,
- know how incorrect initial cross-connections are revealed,
- know why L to cpc values for flat-sheathed cables vary slightly during the test,
- be able to interpret test results

There are two main reasons for conducting this test:

1. To ensure that the ring circuit conductors are continuous, and indicate the value of $(R_1 + R_2)$ for the ring.
2. To establish that interconnections in the ring do not exist.

What then are interconnections in a ring circuit, and why is it important to locate them? Figure 4.1 shows a ring final circuit with an interconnection.

The most likely cause of the situation shown in Figure 4.1 is where a DIY enthusiast has added sockets P, Q, R and S to existing rings A, B, C, D, E and F.

17th Edition IET Wiring Regulations: Inspection, Testing and Certification. 978-1-138-84886-3.

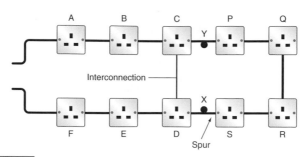

FIGURE 4.1 Ring circuit with an interconnection.

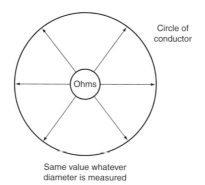

FIGURE 4.2 Measurement across diameter of a circle.

In itself there is nothing wrong with this. The problem arises if a break occurs at, say, point Y, or the terminations fail in socket C or P. Then there would be four sockets all fed from the point X which would then become a spur.

So, how do we identify such a situation with or without breaks at point Y? A simple resistance test between the ends of the line, neutral or circuit protective conductors will only indicate that a circuit exists, whether there are interconnections or not. The following test method is based on the philosophy that the resistance measured across any diameter of a perfect circle of conductor will always be the same value (Figure 4.2).

The perfect circle of conductor is achieved by cross-connecting the line and neutral legs of the ring (Figure 4.3). The test procedure is as follows:

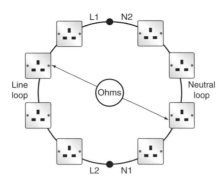

FIGURE 4.3 Measurement across diameter of a ring circuit.

FIGURE 4.4 Ring circuit cross-connections I –N.

1. Identify the opposite legs of the ring. This is quite easy with sheathed cables, but with singles, each conductor will have to be identified, probably by taking resistance measurements between each one and the closest socket outlet. This will give three high readings and three low readings, thus establishing the opposite legs.

2. Take a resistance measurement between the ends of each conductor loop, r_1, r_n and r_2. Record these values.

3. Cross-connect the opposite ends of the line and neutral loops (Figure 4.4).

4. Measure between line and neutral at each socket on the ring. The readings obtained should be, for a perfect ring, substantially the same, and approximately half of the reading of individual loops.

If an interconnection existed such as shown in Figure 4.1, then sockets A–F would all have similar readings, and those beyond the interconnection would have gradually increasing values to approximately the mid point of the ring, then decreasing values back towards the interconnection. If a break had occurred at point Y then the readings from socket S would increase to a maximum at socket P. One or two high readings are likely to indicate either loose connections or spurs. A null reading, that is, an open circuit indication, is probably a reverse polarity, either line- or neutral-cpc reversal. These faults would clearly be rectified and the test at the suspect socket(s) is repeated. If the reading increases dramatically to the centre of the ring and then decreases again, it is likely that incorrect initial cross-connections of the legs of the ring have been made at Step 3.

5. Repeat the above procedure, but in this case cross-connect the line and cpc loops (Figure 4.5).

In this instance, if the cable is of the flat twin type, the readings at each socket will increase very slightly and then decrease around the ring. This difference, due to the line and cpc being different sizes, will not be significant enough to cause any concern. The measured value is very important; it is $R_1 + R_2$ for the ring.

As before, loose connections, spurs and, in this case, L–N cross-polarity will be picked up.

FIGURE 4.5   Ring circuit cross-connections L–cpc.

The details in Table 4.1 are typical approximate ohmic values for a healthy 70 m ring final circuit wired in 2.5 mm²/1.5 mm² flat twin and cpc cable. (In this case the cpc will be approximately 1.673 the L or N resistance.)

As already mentioned, null readings may indicate a reverse polarity. They could also indicate twisted conductors not in their terminal housing. The examples shown in Figure 4.6 may help to explain these situations.

Table 4.1   Resistance Value for a 70 m Ring Circuit

	L1–L2 (r_1)	N1–N2 (r_n)	cpc1–cpc2 (r_2)
Initial measurements	0.52	0.52	0.86
Reading at each socket	0.26	0.26	0.32–0.34
For spurs, each metre in length will add the following resistance to the above values	0.015	0.015	0.02

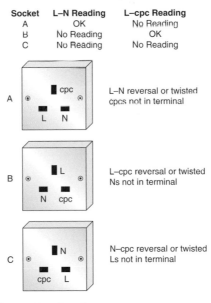

Socket	L–N Reading	L–cpc Reading
A	OK	No Reading
B	No Reading	OK
C	No Reading	No Reading

A — L–N reversal or twisted cpcs not in terminal

B — L–cpc reversal or twisted Ns not in terminal

C — N–cpc reversal or twisted Ls not in terminal

FIGURE 4.6   Reasons for null readings.

Questions

1. State the reasons for conducting a ring final circuit continuity test.
2. What instrument is to be used for the test in Q1 above?
3. Why are interconnections in ring circuits unacceptable?
4. Why are the ends of circuit conductors cross-connected for test purposes?
5. What are the effects on test results of correct and incorrect initial conductor cross-connections?
6. What may a null reading at a socket outlet indicate?
7. What does the L–cpc reading at each socket outlet on a ring signify?
8. Why will the L–cpc readings increase slightly and then decrease around a ring circuit wired in flat sheathed cable?
9. A ring final circuit is wired in 2.5 mm^2 singles (L, N and cpc) in conduit. If each loop has an end-to-end value of 0.4 Ω, what would be the approximate expected value of $(R_1 + R_2)$?

Answers

1. Ensuring the ring is continuous and with no interconnections, and to establish a value for $(R_1 + R_2)$.
2. A low-resistance ohmmeter.
3. A break in the ring beyond an interconnection may leave two or more socket outlets on a spur.
4. To create a perfect circle of conductor, the resistance across any diameter of which will give the same value.
5. Correct cross-connections give the same reading at each socket outlet, incorrect will result in greatly increased and decreased readings around the ring.
6. Twisted or touching conductors not in the socket outlet terminal or a reverse polarity.
7. $(R_1 + R_2)$ for the ring.
8. Because the cpc is smaller in size than the line conductor.
9. 0.2 Ω.

CHAPTER 5

Testing Insulation Resistance (Insulation Resistance Tester)

Important terms/topics covered in this chapter:

- Insulation resistance tester
- Parallel resistances
- Disconnection of equipment
- Test procedure
- Test values
- SELV, PELV and FELV circuits
- Surge protective devices

By the end of this chapter the reader should:

- be aware of why the test is required,
- know the test instrument to be used,
- understand that insulation is a measure of resistances in parallel,
- know between which conductors the measurements should be made,
- know the test voltages and minimum values of insulation resistance,
- be aware of the need to test on circuits/equipment that have been isolated,
- be aware of the reasons for disconnecting various items of equipment,
- be able to calculate overall values of insulation resistance given individual circuit values.

This is probably the most used and yet abused test of them all. Affectionately known as 'meggering', an *insulation resistance test* is performed in order to ensure that the insulation of conductors, accessories and equipment is in a healthy condition, and will prevent dangerous leakage currents between conductors and between conductors and earth. It also indicates whether any short-circuits exist.

Insulation resistance, as just discussed, is the resistance measured between conductors and is made up of countless millions of resistances in parallel (Figure 5.1).

Conductors

Sheathing

Resistance of
installation

FIGURE 5.1   Parallel resistance of cable insulation.

The more resistances there are in parallel, the *lower* the overall resistance, and, in consequence, the longer a cable, the lower the insulation resistance. Add to this the fact that almost all installation circuits are also wired in parallel, and it becomes apparent that tests on large installations may, if measured as a whole, give pessimistically low values, even if there are no faults.

Under these circumstances, it is usual to break down such large installations into smaller sections, floor by floor, distribution circuit by distribution circuit, etc. This also helps, in the case of periodic testing, to minimize disruption. The test procedure is as follows:

1. Ensure the supply to the circuit/s in question is isolated.
2. Disconnect all items of equipment such as capacitors and indicator lamps as these are likely to give misleading results. Remove any

items of equipment likely to be damaged by the test, such as dimmer switches, electronic timers, etc. Remove all lamps and accessories and disconnect fluorescent and discharge fittings. Ensure all fuses are in place, and circuit breakers and switches are in the on position. In some instances it may be impracticable to remove lamps, etc. and in this case the local switch controlling such equipment may be left in the off position. Where electronic devices cannot be disconnected, test only between lives and earth.

3. Join together all live conductors of the supply and test between this join and earth. Alternatively, test between each live conductor and earth in turn.

4. Test between line and neutral. For three phase systems, join together all lines and test between this join and neutral.

5. Then test between each of the lines. Alternatively, test between each of the live conductors in turn. Installations incorporating two-way lighting systems should be tested twice with the two-way switches in alternative positions.

Note: all cpcs should be connected to the earthing arrangement (earth bar) during this test.

Table 5.1 gives the test voltages and minimum values of insulation resistance for ELV and LV systems.

If a value of less than 2 MΩ is recorded it may indicate a situation where a fault is developing, but as yet still complies with the minimum permissible value. In this case each circuit should be tested separately in order to locate the problem.

In the case of SELV, PELV and electrical separation, Table 5.1 applies to their own circuit conductors. When they are with other circuits the insulation resistance between their conductors and those of the other circuits should be based on the highest voltage present. For FELV circuits the test

Table 5.1   Insulation Resistance Test Requirements

System	Test Voltage	Minimum Insulation Resistance
SELV and PELV	250 V d.c.	0.5 MΩ
LV up to 500 V	500 V d.c.	1.0 MΩ
Over 500 V	1000 V d.c.	1.0 MΩ

voltage and the minimum value if insulation is the same as that for LV circuits up to 500 V (i.e. 500 V d.c. and 1 MΩ).

Where surge protective devices exist, they should be disconnected. If this is not practicable the test voltage may be reduced to 250 V d.c. but the minimum value of insulation resistance remains at 1 MΩ.

Example 5.1

An installation comprising six circuits has individual insulation resistances of 2.5, 8, 200, 200, 200 and 200 MΩ, and so the total insulation resistance will be:

$$\frac{1}{R_t} = \frac{1}{2.5} + \frac{1}{8} + \frac{1}{200} + \frac{1}{200} + \frac{1}{200} + \frac{1}{200}$$

$$= 0.4 + 0.125 + 0.005 + 0.005 + 0.005 = 0.545$$

$$\frac{1}{R_t} = \frac{1}{0.545} = 1.83 \text{ MΩ}$$

This is clearly greater than the 1.0 MΩ minimum but less than 2 MΩ. Had this value (1.83) been measured first, the circuits would need to have been investigated to identify the one/s that were suspect.

Note

It is important that a test for cpc continuity is conducted before an insulation resistance (IR) test. If a cpc was broken, and an IR test between line and cpc was carried out first, the result would be satisfactory, even if there was an L–cpc fault beyond the break. A subsequent cpc continuity test would reveal the break, which would be rectified, leaving an L–cpc fault undetected!!

Questions

1. What is the purpose of an insulation resistance test?
2. What instrument should be used?
3. Why do capacitors, neons, etc. need to be disconnected?

4. Why do items of electronic equipment need to be disconnected?
5. What action should be taken regarding switches and protective devices?
6. What is the test voltage and minimum value of insulation resistance for a 25 V FELV circuit?
7. What test voltage and minimum value of insulation resistance are appropriate for circuits incorporating surge protective devices?
8. Below what value of overall insulation resistance would an installation need to be investigated circuit by circuit?
9. Why may a large installation give a pessimistically low overall insulation resistance value?
10. What would be the total insulation resistance of an installation comprising circuits with the following values: 3 MΩ, 12 MΩ, 100 MΩ and 150 MΩ?

Answers

1. To ensure that conductor insulation has not deteriorated or been damaged to an extent that excessive leakage currents can flow.
2. An insulation resistance tester.
3. To avoid misleading test results.
4. To avoid damage to such equipment.
5. All switches ON, all fuses IN, all circuit breakers ON.
6. 500 V d.c.; 1 MΩ.
7. 250 V d.c.; 1 MΩ.
8. 2 MΩ.
9. Because there are a large number of circuits all in parallel.
10. 2.3 MΩ.

CHAPTER 6

Special Tests

The next two tests are special in that they are not often required in the general type of installation. They also require special test equipment. In consequence, the requirements for these tests will only be briefly outlined in this short chapter.

PROTECTION BY BARRIERS OR ENCLOSURES

If, on site, basic protection is provided by fabricating an enclosure or erecting a barrier, it must be shown that the enclosure can provide a degree of protection of at least IPXXB or IP2X or, where required, at least IPXXD or IP4X.

An enclosure having a degree of protection IP2X can withstand the ingress of solid objects exceeding 12 mm diameter and fingers, IPXXB is protection against finger contact only. IP4X gives protection against solid objects and wires exceeding 1 mm in diameter, IPXXD protects against wires exceeding 1 mm in diameter only.

The test for IPXXB or IP2X is conducted with a 'standard test finger' which is supplied at a test voltage not less than 40 V d.c. and not more than 50 V d.c. One end of the finger is connected in series with a lamp and live parts in the enclosure. When the end of the finger is introduced into the enclosure, provided the lamp does not light then the protection is satisfactory (Figure 6.1).

The test for IPXXD or IP4X is conducted with a rigid 1 mm diameter wire with its end cut at right angles. Protection is afforded if the wire does not enter the enclosure.

PROTECTION BY NON-CONDUCTING LOCATION

This is a rare location and demands specialist equipment to measure the insulation resistance between insulated floors and walls at various points. Appendix 13 of BS 7671 outlines the tests required.

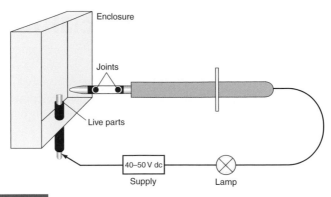

FIGURE 6.1   BS finger test.

CHAPTER 7

Testing Polarity (Low-Resistance Ohmmeter)

Important terms/topics covered in this chapter:

- Edison screw lampholders
- Radial socket outlet circuits
- Supply polarity

By the end of this chapter the reader should:

- know the instrument to be used,
- know why BS EN 60238, E14 and E27 lampholders are exempt from polarity testing,
- know why ring final circuit polarity is not usually carried out during polarity testing,
- know how to check for line–cpc reversals on radial socket outlet circuits,
- know what live polarity test should be conducted.

This simple test, often overlooked, is just as important as all the others, and many serious injuries and electrocutions could have been prevented if only polarity checks had been carried out.

The requirements are:

1. All fuses and single pole switches and protective devices are in the line conductor.
2. The centre contact of an Edison screw type lampholder is connected to the line conductor (except E14 and 27 types to BS EN 60238, as these have threads of insulating material and the lamp must be fully inserted before L and N contacts are made).
3. All socket outlets and similar accessories are correctly wired.

Although polarity is towards the end of the recommended test sequence, it would seem sensible, on lighting circuits, for example, to conduct this test at the same time as that for continuity of cpcs (Figure 7.1).

17th Edition IET Wiring Regulations: Inspection, Testing and Certification. 978-1-138-84886-3.
© Brian Scaddan. Published by Taylor & Francis. All rights reserved.

As discussed earlier, polarity on ring final circuit conductors is achieved simply by conducting the ring circuit test. For radial socket outlet circuits, however, this is a little more difficult. The continuity of the cpc will have already been proved by linking line and cpc and measuring between the same terminals at each socket. Whilst a line–cpc reversal would not have shown, a line–neutral reversal would, as there would have been no reading at the socket in question. This would have been remedied, and so only line–cpc reversals need to be checked. This can be done by linking together cpc and neutral at the origin and testing between the same terminals at each socket. A line–cpc reversal will result in no reading at the socket in question.

For lighting circuits, the test is the same as the $R_1 + R_2$ test, so polarity is checked then. The same applies to the radial socket outlet circuits if the socket fronts are open to test at the actual cable terminations, as line–cpc reversals will be visible.

Live polarity

When the supply is connected, it is important to check that the incoming supply is correct. This is done using an approved voltage indicator at the intake position or close to it.

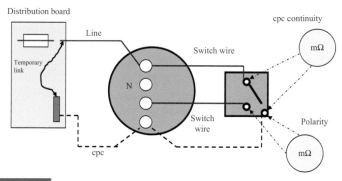

FIGURE 7.1 Lighting circuit polarity.

Questions

1. What instrument is used for testing polarity?
2. Why are BS EN 60238 E14 and E27 lampholders exempt from polarity testing?
3. At what point in a test sequence is the polarity of a ring final circuit checked?
4. How are line–cpc reversals identified in radial socket outlet circuits?
5. Where should live polarity tests be conducted?

Answers

1. Low-resistance ohmmeter.
2. The lampholder screw thread is made of an insulating material.
3. When the ring final circuit continuity test is being conducted.
4. By cross-connecting neutral and cpc and testing between N and cpc at each socket.
5. At the supply intake to the installation.

CHAPTER 8

Testing Earth Electrode Resistance (Earth Electrode Resistance Tester or Loop Impedance Testers)

Important terms/topics covered in this chapter:

- Earth electrode resistance area
- Potential divider
- Current and potential electrodes
- Average value of earth electrode resistance
- Use of earth fault loop impedance tester

By the end of this chapter the reader should:

- know the test instruments that may be used,
- understand what is meant by the resistance area of an earth electrode,
- be able to state the electrodes involved when using an earth electrode resistance tester,
- know the extent of the resistance area of an electrode,
- know how to conduct a test using an earth electrode resistance tester,
- know what test may be conducted when the system is TT and is RCD protected,
- be able to determine the value of earth electrode resistance from test results.

In many rural areas, the supply system is TT and hence reliance is placed on the general mass of earth for a return path under earth fault conditions. Connection to earth is made by an electrode, usually of the rod type, and preferably installed as shown in Figure 8.1.

In order to determine the resistance of the earth return path, it is necessary to measure the resistance that the electrode has with earth. If we were to make such measurements at increasingly longer distances from

FIGURE 8.1 Earth electrode installation.

the electrode, we would notice an increase in resistance of up to about 2.5–3 m from the rod, after which no further increase in resistance would be noticed (Figure 8.2).

The maximum resistance recorded is the electrode resistance and the area that extends to 2.5–3 m beyond the electrode is known as the earth electrode resistance area.

There are two methods of making the measurement, one using a proprietary instrument and the other using a loop impedance tester.

METHOD 1: PROTECTION BY OVERCURRENT DEVICE

This method is based on the principle of the potential divider (Figure 8.3).

By varying the position of the slider the resistance at any point may be calculated from $R = V/I$.

The earth electrode resistance test is conducted in a similar fashion with the earth replacing the resistance and a potential electrode replacing the slider (Figure 8.4). In Figure 8.4, the earthing conductor to the electrode under test is temporarily disconnected.

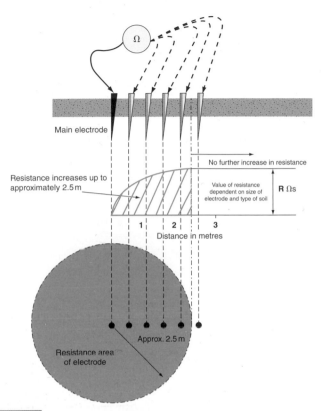

FIGURE 8.2 Earth electrode resistance area.

FIGURE 8.3 Potential divider.

FIGURE 8.4   Earth electrode resistance test.

The method of test is as follows:

1. Place the current electrode (C2) away from the electrode under test, approximately 10 times its length (i.e. 30 m for a 3 m rod).
2. Place the potential electrode midway.
3. Connect test instrument as shown.
4. Record resistance value.
5. Move the potential electrode approximately 6 m either side of the mid position, and record these two readings.
6. Take an average of these three readings (this is the earth electrode resistance).

For TT systems the result of this test will indicate compliance if the product of the electrode resistance and the operating current of the overcurrent device does not exceed 50 V. Clearly this will not be achieved when electrode resistances are high and hence will be more appropriate for electrodes used for earth connections for transformers and generators where the values need to be very small. Generally speaking the values obtained will result in the need for RCD protection.

METHOD 2: PROTECTION BY A RESIDUAL CURRENT DEVICE

In this case, an earth fault loop impedance test is carried out between the incoming line terminal and the electrode (a standard test for Z_e).

The value obtained is added to the cpc resistance of the protected circuits and this value is multiplied by the operating current of the RCD. The resulting value should not exceed 50 V. If it does, then Method 1 should be used to check the actual value of the electrode resistance.

Questions

1. What instruments may be used for earth electrode resistance testing?
2. What is the extent of the resistance area of an earth electrode?
3. For a 4 m electrode under test, at what distance away should the current electrode be placed?
4. Where should a potential electrode be initially placed when conducting an earth electrode resistance test?
5. Where are the alternative positions for the potential electrode?
6. What would be the resistance of an earth electrode if the test results gave values of 127 Ω, 129 Ω and 122 Ω?
7. What test may be performed when the system is TT and protected by an RCD?

Answers

1. Earth electrode resistance tester or earth fault loop impedance tester.
2. Approximately 2.5 m radius from the electrode.
3. 40 m minimum.
4. Centrally between the electrode under test and the current electrode.
5. 6 m either side of the potential electrode's initial position.
6. 126 Ω.
7. An earth fault loop impedance test.

CHAPTER 9

Testing Earth Fault Loop Impedance Tester

Important terms/topics covered in this chapter:

- Earth fault loop path
- Comparison of results with maximum values
- The rule of thumb
- RCD and cb operation
- Calculation of loop impedance
- External earth fault loop impedance

By the end of this chapter the reader should:

- know what instrument is required,
- be conversant with the various earth fault loop paths,
- know the test procedure,
- know how to adjust maximum values for comparison with test values,
- know to overcome the problems of RCD or cb operation during the test,
- be aware of the requirements for testing external earth fault loop impedance.

This is very important but, sadly, poorly understood. So let us remind ourselves of the component parts of the earth fault loop path (Figure 9.1). Starting at the point of fault:

1. The cpc.
2. The earthing conductor and main earthing terminal.
3. The return path via the earth for TT systems, and the metallic return path in the case of TN-S or TN-C-S systems. In the latter case the metallic return is the PEN conductor.
4. The earthed neutral of the supply transformer.
5. The transformer winding.
6. The line conductor back to the point of fault.

17th Edition IET Wiring Regulations: Inspection, Testing and Certification. 978-1-138-84886-3.
© Brian Scaddan. Published by Taylor & Francis. All rights reserved.

FIGURE 9.1 Earth fault loop path.

Overcurrent protective devices must, under earth fault conditions, disconnect fast enough to reduce the risk of electric shock. This is achieved if the actual value of the earth fault loop impedance does not exceed the tabulated maximum values given in the relevant parts of the IET Regulations.

The purpose of the test, therefore, is to determine the actual value of the loop impedance (Z_s), for comparison with those maximum values, and it is conducted as follows:

1. Ensure that all main equipotential bonding is in place.
2. Connect the test instrument either by its BS 1363 plug, or the 'flying leads', to the line, neutral and earth terminals at the remote end of the circuit under test. (If a neutral is not available, e.g. in the case of a three-phase motor, connect the neutral probe to earth.)
3. Press to test and record the value indicated.

It must be understood that this instrument reading is *not valid for direct comparison with the tabulated maximum values*, as account must be taken of the ambient temperature at the time of test and the maximum conductor operating temperature, both of which will have an effect on conductor resistance. Hence, the $(R_1 + R_2)$ could be greater at the time of fault than at the time of test.

So, our measured value of Z_s must be corrected to allow for these possible increases in temperature occurring at a later date. This requires actually measuring the ambient temperature and applying factors in a formula.

Clearly this method of correcting Z_s is time consuming and unlikely to be commonly used. Hence, a rule of thumb method may be applied which simply requires that the measured value of Z_s does not exceed 0.8 of the appropriate tabulated maximum value. Table 9.1 gives the 0.8 values of tabulated loop impedance for direct comparison with measured values.

In effect, a loop impedance test places a line/earth fault on the installation, and if an RCD is present it may not be possible to conduct the test as it will keep tripping out. Unless the instrument can compensate for this, the value of Z_s will have to be calculated using the measured values of Z_e and $(R_1 + R_2)$ and the 0.8 rule applied. Remember, $Z_s = Z_e + (R_1 + R_2)$.

EXTERNAL LOOP IMPEDANCE Z_e

The value of Z_e is measured at the origin of the installation on the supply side with the means of earthing disconnected, to avoid parallel paths. Do not conduct this test if the installation cannot be isolated.

Important Note

Never bypass an RCD in order to conduct this test. Also, as this test creates a high current, some lower rated cbs may operate on overload. Do not replace with a higher rated breaker for test purposes; use the calculation method.

Table 3.5 Corrected Maximum Z_s Values for Comparison with Measured Values

PROTECTIVE DEVICE		PROTECTION RATING

PROTECTIVE DEVICE		5	6	10	15	16	20	25	30	32	40	45	50	60	63	80	100	125	160	200
BS EN 60898 and 61009 type **B**	0.4s & 5s		5.82	3.5		2.19	1.75	1.4		1.09	0.87		0.7		0.56	0.5	0.35	0.29		
BS EN 60898 and 61009 type **C**	0.4s & 5s		2.9	1.75		1.09	0.87	0.71		0.54	0.44		0.35		0.28	0.2	0.17	0.14		
BS EN 60898 and 61009 type **D**	0.4s		1.46	0.87		0.54	0.44	0.35		0.28	0.22		0.17		0.13	0.1	0.08	0.07		
	5s		2.91	1.75		1.09	0.87	0.69		0.55	0.44		0.35		0.27	0.22	0.17	0.14		
BS 3036 Semi-enclosed	0.4s	7.22			1.94		1.34		0.83											
	5s	13.45			4.1		2.9		2			1.2		0.85			0.4			
		3A*			13A*															
BS 1362 Cartridge	0.4s	12.5			1.8															
	5s	17.63			2.9															

BS 38-2 Bolted type E and Clip-in type G

0.4s		6.24	3.71		1.95	1.35	1.03		0.79				
5s		9.73	5.46	3.17	2.24	1.75	1.4	1.03	0.79	0.64	0.4	0.35	0.26	0.21	0.2

BS 38-3 Cartridge type C

0.4s	7.94			1.84	1.55		0.73				
5s	11.62			2.14	2.57	1.24	0.79	0.55	0.4	0.3	

(Continued)

Table 3.5 Corrected Maximum Z_s Values for Comparison with Measured Values (*Continued*)

OBSOLETE FUSES AND CIRCUIT BREAKERS

MAXIMUM CORRECTED Z_s VALUES

PROTECTION RATING

		5	6	10	15	16	20	25	30	32	40	45	50	60	63	80	100	125	160	200
BS 88 – 2.2 and 88-6	0.4s		6.48	3.89		2.05	1.35	1.09		0.79										
	5s		10.26	5.64		3.16	2.2	1.75		1.4	1.03		0.79		0.64	0.5	0.32	0.25	0.19	0.2
BS 1361 Cartridge	0.4s	7.94			2.45		1.29		0.87											
	5s	12.46			3.8		2.13		1.4			0.75		0.53		0.4	0.28			
BS 3871 type 1	0.4s & 5s	8.74	7.22	4.37	2.9	2.73	2.19	1.75	1.45	1.37	1.09	0.97	0.87		0.69					
BS 3871 type 2	0.4s & 5s	5	4.15	2.49	1.66	1.56	1.24	1	0.83	0.78	0.64	0.55	0.49		0.42					
BS 3871 type 3	0.4s & 5s	3.5	2.85	1.75	1.16	1.09	0.87	0.7	0.58	0.54	0.44	0.39	0.35		0.28					

Questions

1. What instrument is used for earth fault loop impedance testing?
2. Which earthing system includes a PEN conductor?
3. Before testing, what action should be taken regarding equipotential bonding?
4. Why is the 0.8 rule applied?
5. Is a measured value of loop impedance of 1.2 Ω satisfactory if the tabulated maximum value is 1.44 Ω?
6. How may a value for loop impedance Z_s be obtained if an RCD or a cb operates when the test is conducted?
7. What action is required regarding the earthing conductor of an installation before conducting a test for external loop impedance Z_e?
8. Why is the action in Q7 above required and what other measure must be taken?

Answers

1. An earth fault loop impedance tester.
2. TN-C-S.
3. Ensure it is connected.
4. To compensate for increased ambient and conductor operating temperature.
5. No, as the corrected maximum would be 0.8 × 1.44 = 1.15 Ω.
6. Calculation from $Z_s = Z_e + (R_1 + R_2)$.
7. It must be disconnected.
8. To avoid parallel paths. The supply to the installation must be isolated.

CHAPTER 10

Additional Protection (RCD Tester)

Important terms/topics covered by this chapter:

- RCD/RCBO test requirements
- Uses for RCDs/RCBOs
- Determination of RCD/RCBO rating

By the end of this chapter the reader should:

- know what instrument should be used,
- know the test requirements for various types of RCD/RCBO,
- know the instrument settings required,
- be able to identify where RCDs/RCBOs are required,
- know how to determine the rating of RCDs/RCBOs.

RCD/RCBO OPERATION

Where RCDs and RCBOs are used as additional protection against shock, it is essential that they operate within set parameters. The RCD testers used are designed to do just this, and the basic tests required are as follows (Table 10.1):

1. Set the test instrument to the rating of the RCD.
2. Set the test instrument to half-rated trip ($1/2\ I_{\Delta n}$).
3. Operate the instrument and the RCD should not trip.
4. Set the instrument to deliver the full-rated tripping current of the RCD ($I_{\Delta n}$).

Note

- A loop impedance test must be conducted before the RCD test as high values of loop impedance or the absence of an earth path could result in dangerous voltages appearing between exposed and extraneous conductive parts and persons should be told not to touch metalwork whilst the test is being conducted.
- Most RCD testers have the facility to test, separately, each half cycle of the supply and so each test should be done at 0° and 180°. The highest reading should be recorded

17th Edition IET Wiring Regulations: Inspection, Testing and Certification. 978-1-138-84886-3.

5. Operate the instrument and the RCD should trip out in the required time.
6. A 30 mA RCD or less, operating at $5 \times I_{\Delta n}$, should trip in 40 ms.

Table 10.1 RCD/RCBO Test Requirements

RCD Type	Half-Rated	Full-Rated Trip Current
BS 4239 and BS 7288 sockets	No trip	<200 ms
BS 4239 with time delay	No trip	½ time delay + 200 ms to time delay + 200 ms
BS EN 61009 or BS EN 61009 RCBO	No trip	<300 ms
As above but Type S with time delay	No trip	130 ms ≤ I ≤ 500 ms

Note

This last test is not required for RCDs rated over 30 mA.

There seems to be a popular misconception regarding the ratings and uses of RCDs in that they are the panacea for all electrical ills and the only useful rating is 30 mA!

First, RCDs are not fail-safe devices; they are electromechanical in operation and can malfunction. Second, general purpose RCDs are manufactured in ratings from 5 to 500 mA and have many uses. The accepted lethal level of shock current is 50 mA and hence RCDs rated at 30 mA or less would be appropriate for use where shock is an increased risk. The following list indicates the residual current ratings and uses of RCDs as stated in BS 7671.

REQUIREMENTS FOR RCD PROTECTION

30 mA

- All socket outlets rated at not more than 20 A and for unsupervised general use.
- Mobile equipment rated at not more than 32 A for use outdoors.
- All circuits in a bath/shower room.

- Preferred for all circuits in a TT system.
- All cables installed less than 50 mm from the surface of a wall or partition in prescribed zones and not protected by any other method, and also at any depth if the construction of the wall or partition includes metallic parts.
- In zones 0, 1 and 2 of swimming pool locations.
- All circuits in a location containing saunas, etc.
- Socket outlet final circuits not exceeding 32 A in agricultural locations.
- Circuits supplying Class II equipment in restrictive conductive locations.
- Each socket outlet in caravan parks and marinas and final circuit for houseboats.
- All socket outlet circuits rated not more than 32 A for show stands, etc.
- All socket outlet circuits rated not more than 32 A for construction sites (where reduced low voltage, etc. is not used).
- All socket outlets supplying equipment outside mobile or transportable units.
- All circuits in caravans.
- All circuits in circuses, etc.
- A circuit supplying Class II heating equipment for floor and ceiling heating systems.

100 mA

- Socket outlet final circuits of rating exceeding 32 A in agricultural locations.

300 mA

- At the origin of a temporary supply to circuses, etc.
- Where there is a risk of fire due to storage of combustible materials.
- All circuits (except socket outlets) in agricultural locations.

500 mA

- Any circuit supplying one or more socket outlets of rating exceeding 32 A, on a construction site.

Where loop impedance values cannot be met, RCDs of an appropriate rating can be installed. Their rating can be determined from

$$I_{\Delta n} = \frac{50}{Z_s}$$

where $I_{\Delta n}$ is the rated operating current of the device, 50 is the touch voltage, and Z_s is the measured loop impedance.

Questions

1. What test instrument is required for RCD/RCBO testing?
2. What is the maximum operating time for a BS EN 61008 RCD at full rated current?
3. What is the maximum operating time for a 30 mA RCD when tested at 150 mA?
4. What maximum rating of RCD should be used for a 63 A socket outlet on a construction site?
5. What rating of RCD is required for a caravan installation?
6. What would be the required maximum rating of an RCD where the earth fault loop impedance was 167 Ω?

Answers

1. An RCD tester.
2. 300 ms.
3. 40 ms.
4. 500 mA.
5. 30 mA.
6. 300 mA.

CHAPTER 11

Prospective Fault Current (PFC/PSCC Tester)

There is a requirement to determine the prospective fault current at the origin of an installation and at relevant points throughout. At the origin this may be ascertained by enquiry or measurement, whereas at other points measurement is the only option.

Where the lowest rated protective device in the installation has a breaking capacity higher than the PFC at the origin, then measurement at other points is not needed.

The purpose of this test is to verify that the breaking capacity of a protective device is suitable for the prospective fault current at the point that it is installed.

A protective device with a breaking capacity of, say, 3 kA installed in a distribution board where the PFC is 6 kA could result in serious damage and/or fire in the event of a fault.

BS EN 60898 circuit breakers have two values of breaking capacity, the service rating Ics and the fault rating Icn. The Icn rating is marked on the device, e.g. 3000; at this value of fault current the breaker will operate but may not be usable afterwards.

PFC is a generic term and can be either prospective short-circuit current, PSCC (between lines or line and neutral) or prospective earth fault current, PEFC (between line and earth). Both should be measured and the highest value recorded, although there is no harm in recording both.

The testers are designed for single-phase use, so where a value of PSCC is required for a three-phase system it may be determined by multiplying the single phase by 2, or more accurately 1.732.

17th Edition IET Wiring Regulations: Inspection, Testing and Certification. 978-1-138-84886-3.

CHAPTER 12

Check of Phase Sequence (Phase Sequence Indicator)

For three-phase systems it is important to have knowledge of the phase rotation of the supply and at various points within an installation. It is convention for this rotation to be normally Brown, Black, Grey, or LI, L2, L3. The direction of three-phase motors can be reversed simply by reversing any two phases. In consequence the correct sequence is essential to ensure the right rotation.

Paralleling of two three-phase generators or of a generator to the three-phase public supply system requires their phase sequences to be synchronized.

The instrument is simply a small three-phase motor with a dial that indicates in which direction the supply is rotating (Figure 12.1).

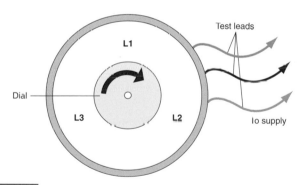

FIGURE 12.1   Phase sequence indicator.

CHAPTER 13

Functional Testing

All RCDs have a built-in test facility in the form of a test button. Operating this test facility creates an artificial out-of-balance condition that causes the device to trip. This only checks the mechanics of the tripping operation; it does not check the condition of the electrical installation and hence is not a substitute for the tests discussed in Chapter 10.

There should be a notice in a prominent position at or near the origin of the installation where the device is located indicating that the test button should be operated quarterly. For temporary installations it is recommended that this operation be carried out at more regular intervals.

All other items of equipment such as switchgear, controlgear interlocks, etc. must be checked to ensure that they are correctly mounted and adjusted, and that they function correctly.

This could involve the operation of, for example, two-way switching, dimmer switches, timers, thermostats, main isolators, circuit breakers, etc.

CHAPTER 14

Voltage Drop (Approved Voltmeter)

There may be a requirement to determine the voltage at the terminals of equipment to ensure that items of electrical equipment will function correctly and safely. As the maximum volt drop will occur when the circuit is at full load and at maximum cable operating temperature it would be unlikely that an accurate instrument measurement would be achievable. Hence the calculation method is used.

This is simply based on good old Ohm's law: $V = I \times R$.

All we need is the length and csa of the cable/conductors or a measured value of resistance and the full load current.

Example 14.1

A motor with a full load current of 10 A is fed from a 230 V supply by a 70°C thermoplastic cable of total resistance (L+N) 0.222 Ω measured at 20°C. What is the expected voltage drop and is it acceptable?

Volt drop $V = I \times R = 10 \times 0.222 \times 1.2 = 2.66$ V (the 1.2 factor converts the 20°C resistance to what it would be at 70°C). This is acceptable, as the maximum is 11.5 V.

Alternatively we could use the length, the csa and the tabulated resistance in mΩ/m at 20°C.

In the example above the length would have been 15 m and the single conductor csa 2.5 mm² which has a resistance of 7.41 mΩ at 20°C.

$$\text{Volt drop } V = I \times \frac{(L \times 2 \times m\Omega/m \times 1.2)}{1000}$$

$$= \frac{10 \times 15 \times 2 \times 7.41 \times 1.2}{1000} = 2.66 \text{ V}$$

17th Edition IET Wiring Regulations: Inspection, Testing and Certification. 978-1-138-84886-3.

Table 14.1 Maximum Permissible Voltage Drop		
	LV Lighting 3%	**LV Power 5%**
Single-phase 230 V	6.9 V	11.5 V
Three-phase 400 V	12 V	20 V

CHAPTER 15

Periodic Inspection

Important terms/topics covered in this chapter:

- Circumstances requiring periodic inspection and testing
- General reasons for periodic inspection and testing
- Conditions to be investigated
- Documentation to be completed
- General areas of non-compliance that may be revealed

By the end of this chapter the reader should:

- know why periodic inspection and testing is required,
- know what general conditions within an installation need investigation,
- know what documentation needs to be completed,
- know what action is required if there is insufficient information/ drawings, etc.,
- understand the extent to which dismantling and sampling should take place,
- be aware of the conditions that may permit an installation to be exempt from periodic inspection and testing.

PERIODIC INSPECTION AND TESTING

This is the province of the experienced inspector who has not only the knowledge and technical expertise to competently carry out the testing process but who is also fully conversant with correct electrical installation practices.

The purpose of the periodic inspection and testing is to establish whether or not the installation is in a satisfactory condition for continued use. This is achieved by relevant inspection and appropriate testing. The most important test is the loop impedance test, as this will establish if disconnection times are being met.

There are many instances where testing is not required. For example, the main purpose of conducting a ring circuit continuity test is to establish if there are interconnections in the ring. Thus, if there are records of the test results from previous inspections and there has been no work carried out on the circuit, then it is impossible for an interconnection to happen on its own! Hence the ring circuit test is a waste of time.

In fact, even $R_1 + R_2$ tests can be avoided as the Z_s tests will show the continuity of cpcs. The only other test which should be carried out is insulation resistance, as there may have been some deterioration since the last inspection.

Clearly, if there are no previous test results, more comprehensive testing will be required.

In any case the inspector should conduct a walk round to establish the extent of any sampling of the circuits to be inspected. This sampling may be reduced or increased as the inspection proceeds, depending on the findings.

It should also be noted that in its true sense, Periodic Inspection and Testing does not include fault finding as this is time consuming and financially prohibitive. Remember, the purpose is to report on the condition of the installation, not to rectify it.

CIRCUMSTANCES WHICH REQUIRE A PERIODIC INSPECTION AND TEST

Test and inspection is due; insurance, mortgage, licensing reasons; change of use; change of ownership; after additions or alterations; after damage; change of loading; to assess compliance with current regulations.

GENERAL REASONS FOR A PERIODIC INSPECTION AND TEST

1. To ensure the safety of persons and livestock.
2. To ensure protection of property from fire and heat.
3. To ensure that the installation is not damaged so as to impair safety.
4. To ensure that the installation is not defective and complies with the current regulations.

GENERAL AREAS OF INVESTIGATION

Safety, Ageing, Damage, Corrosion, Overloading, Wear and tear, External influences and Suitability (as an *aide memoire* the author calls this his SADCOWES list!).

DOCUMENTATION TO BE COMPLETED

An electrical installation condition report (EICR), a schedule of test results and a schedule of inspections.

SEQUENCE OF TESTS

There is no required sequence. However, the sequence for an Initial Verification is preferred if possible.

Periodic inspection and testing could be such a simple and straightforward process. Nevertheless, it usually tends to be complicated and frustrating.

On the domestic scene, I doubt if any house owner actually decides to have a regular inspection, the comment being, 'If it works it must be OK'. It is usually only when there is a change of ownership that the mortgage companies insist on an electrical survey. The worst cases are, however, in industry and commerce.

Periodic inspections are often requested by clients, reluctantly, to satisfy insurers or an impending visit by the HSE. Even then it is usually the case that 'you can't turn that off' or 'why can't you just test this bit and then issue a certificate for the whole lot'.

Under the rare circumstances that an inspection and test is genuinely requested due to responsible concerns for the safety of staff, etc., it is difficult to convince the client that, as there are no drawings, or information about the installation, and that no switchgear is labelled, etc., you are going to be on site for a considerable time and at a considerable cost.

When there are no drawings or items of information, especially on a large installation, there may be a degree of exploratory work to be carried out in order to ensure safety whilst inspecting and testing. If it is felt that it may be unsafe to continue with the inspection and test, then drawings and information must be produced in order to avoid contravening Section 6 of the Health and Safety at Work Act.

However, let us assume (in our wildest dreams) that, as with an Initial Verification, the original installation was erected in accordance with the Wiring Regulations, and any alterations and/or additions have been faithfully recorded and all the original documentation/diagrams/charts, etc. are readily available!

A periodic inspection and test under these circumstances should be relatively easy, as little dismantling of the installation will be necessary, and the bulk of the work will be inspection.

Inspection should be carried out with the supply disconnected as it may be necessary to gain access to wiring in enclosures, etc., and hence with large installations it will probably need considerable liaison with the client to arrange convenient times for interruption of supplies to various parts of the installation.

This is also the case when testing protective conductors, as these must *never* be disconnected unless the supply can be isolated. It is particularly important in the case of main protective bonding conductors which need to be disconnected in order to measure Z_e.

In general an inspection should reveal:

1. Any aspects of the installation that may impair the safety of persons and livestock against the effects of electric shock and burns.
2. That there are no installation defects that could give rise to heat and fire, and hence damage property.
3. That the installation is not damaged or deteriorated so as to impair safety.
4. That any defects or non-compliance with the Regulations, which may give rise to danger, are identified.

As was mentioned earlier, dismantling should be kept to a minimum as this process may create faults. Hence a certain amount of sampling will be required.

The amount of sampling would need to be commensurate with the number of defects being found.

It should be noted that if an installation is effectively supervised in normal use, then periodic inspection and testing can be replaced by regular maintenance by skilled persons. This would only apply to, say, factory installations where there are permanent maintenance staff.

Questions

1. State two circumstances that would result in the need for a periodic inspection.
2. State three installation conditions that may need investigation.
3. State the three items of documentation that will need to be completed.
4. When may exploratory work be required before commencing a periodic inspection and test of a large installation?
5. Why should dismantling be kept to a minimum?
6. When may periodic inspection and testing be replaced by routine maintenance?

Answers

1. Any two, from due date, mortgage, insurance, etc.
2. Any three from SADCOWES list.
3. Periodic/condition report.
4. Where there is a lack of information/drawings, etc., and it may be unsafe to continue without them.
5. To avoid causing damage and creating faults.
6. When the installation is under effective supervision and the maintenance is carried out by skilled persons.

CHAPTER 16

Certification

Having completed all the inspection checks and carried out all the relevant tests, it remains to document all this information. This is done on Electrical Installation Certificates, Electrical Installation Condition Reports, schedules, test results, Minor Electrical Installation Works Certificates and any other documentation you wish to append to the foregoing. Examples of such documentation are shown in BS 7671 and the IET Guidance Note 3 on inspection and testing.

This documentation is vitally important. It has to be correct and signed or authenticated by a competent person. Electrical Installation Certificates and Electrical Installation Condition Reports must be accompanied by a schedule of test results and a schedule of inspections for them to be valid. It should be noted that both Electrical Installation Certificates and Minor Electrical Installation Works Certificates should be signed or otherwise authenticated by competent persons in respect of the design, the construction and the inspection and testing of the installation. The Electrical Installation Condition Report is signed by the inspector.

(For larger installations there may be more than one designer, hence the certificate has space for two signatures, i.e. designer 1 and designer 2.) It could be, of course, that for a very small company, one person signs all three parts. Whatever the case, the original must be given to the person ordering the work, and a duplicate retained by the contractor.

One important aspect of an EIC is the recommended interval between inspections. This should be evaluated by the designer and will depend on the type of installation and its usage. In some cases the time interval is mandatory, especially where environments are subject to use by the public. The IET Guidance Note 3 give recommended maximum frequencies between inspections.

An Electrical Installation Condition Report is very similar in part to an Electrical Installation Certificate in respect of details of the installation, i.e. maximum demand, type of earthing system, Z_e, etc. The rest of the form deals with the extent and limitations of the inspection and test, recommendations, and a summary of the installation. The record of the extent and limitations of the inspection is very important. It must be agreed with the client or other third party exactly what parts of the installation will be covered by the report and those that will not. The interval until the next test is determined by the inspector.

With regard to the schedule of test results, test values should be recorded unadjusted, any compensation for temperature, etc. being made after the testing is completed.

Any alterations or additions to an installation will be subject to the issue of an Electrical Installation Certificate, except where the addition is, say, a single point added to an existing circuit, when the work is subject to the issue of an MEIWC.

Summary:

1. The addition of points to existing circuits requires a Minor Electrical Installation Works Certificate.
2. A new installation or an addition or alteration that comprises new circuits requires an Electrical Installation Certificate.
3. An existing installation requires an Electrical Installation Condition Report.

Note

Points (2) and (3) must be accompanied by a schedule of test results and a schedule of inspections.

As the client/customer is to receive the originals of any certification, it is important that *all* relevant details are completed correctly. This ensures that future inspectors are aware of the installation details and test results which may indicate a slow progressive deterioration in some or all of the installation.

These certificates, etc. could also form part of a 'seller's pack' when a client wishes to sell a property.

The following is a general guide to completing the necessary documentation and should be read in conjunction with the examples given in BS 7671 and the IET Guidance Note 3.

ELECTRICAL INSTALLATION CERTIFICATE

1. Details of client:
   Name: *Full name.*
   Address: *Full address and postcode.*
   Description: *Domestic, industrial, commercial.*
   Extent: *What work has been carried out (e.g. full rewire, new shower circuit, etc.). Tick relevant box.*
2. Designer/constructor/tester:
   Details of each or could be one person.
   Note: Departures are not faults, they are systems/equipment, etc. that are not detailed in BS 7671 but may be perfectly satisfactory.
3. Next test:
   When the next test should be carried out and decided by the designer.
4. Supply characteristics and earthing arrangements:
   Earthing system: *Tick relevant box (TT, TN-S, etc.).*
   Live conductors: *Tick relevant boxes.*
   Nominal voltage: *Obtain from supplier, but usually 230 V single-phase U and U_0 but 400 V U and 230 U_0 for three-phase.*
   Frequency: *From supplier but usually 50 Hz.*
   PFC: *From supplier or measured. Supplier usually gives 16 kA.*
   Z_e: *From supplier or measurement. Supplier usually gives 0.8 V for TN-S; 0.35 V for TN-C-S and 21 V for TT systems.*
   Main fuse: *Usually BS 1361, rating depends on maximum demand.*
5. Particulars of installation:
   Means of earthing: *Tick 'supplier's facility' for TN systems, 'earth electrode' for TT systems.*
   Maximum demand: *Value without diversity.*
   Earth electrode: *Measured value or N/A.*
   Earthing and bonding:
   Conductors: *Actual sizes and material, usually copper.*

Main switch or circuit breaker (could be separate units or part of a consumer control unit): *BS number; Rating, current and voltage; Location; 'not address'; that is, where it is located in the building; Fuse rating if in a switch-fuse, else N/A; RCD details only if used as a main switch.*

6. Comments on existing installation:

Write down any defects found in other parts of the installation which may have been revealed during an addition or an alteration.

7. Schedules:

Indicate the number of test and inspection 'schedules that will accompany this certificate'.

ELECTRICAL INSTALLATION CONDITION REPORT

1. Details of client:

Name: *Full name (could be a landlord, etc.).*

Address: *Full address and postcode (may be different to the installation address).*

Purpose: *For example, due date; change of owner/tenant; change of use, etc.*

2. Details of installation:

Occupier: *Could be the client or a tenant.*

Installation: *Could be the whole or part (give details).*

Address: *Full and postcode.*

Description: *Tick relevant box.*

Age: *If not known, say so, or educated guess.*

Alterations: *Tick relevant box and insert age where known.*

Last inspection: *Insert date or 'not known'.*

Records: *Tick relevant box.*

3. Extent and limitations:

Full details of what is being tested (extent) and what is not (limitations). If not enough space on form add extra sheets.

4. Next inspection:

Filled in by inspector and signed, etc. under declaration.

5. Supply details:

As per an Electrical Installation Certificate.

6. Observations:

Tick relevant box, if work is required, record details and enter relevant code (C1, C2, C3 and F1) in space on right-hand side.

7. Summary:

Comment on overall condition. Only the identification of a code C1 or C2 will require an installation to be classified as being UNSATISFACTORY.

8. Schedules:

Attach completed schedules of inspections and test results.

OBSERVATION CODES

C1 Danger present, immediate remedial action required.
This is used where there is, for example, an exposed live part that can be touched. Its is suggested that if practicable, this be rectified on discovery but, if not, then the owner must be informed in writing as a matter of urgency.

C2 Potentially dangerous, urgent remedial action required.
This would be the case, say, where main protective bonding was not present. It would only become dangerous if there was a fault to earth and a person happened to be simultaneously touching accessible exposed and extraneous conductive parts.

C3 Improvement recommended.
This would be used, for example, if there was a label missing for non-standard colours. It is an omission but it does not make the installation unsafe for the user.

F1 Further investigation needed without delay.
This would be the case when a code C1 or C2 defect is suspected but because of extent and limitations the circuit/s in question cannot be accessed.

MINOR ELECTRICAL INSTALLATION WORKS CERTIFICATE

Only to be used when simple additions or alterations are made, *not when a new circuit is added.*

1. Description: *Full description of work.*
Address: *Full address.*

Date: *Date when work was carried out.*

Departures: *These are not faults, they are systems/equipment, etc. that are not detailed in BS 7671 but may be perfectly satisfactory (this is usually N/A).*

2. Installation details:

   Earthing: *Tick relevant box.*

   Method of fault protection: *99% of the time this will be automatic disconnection of supply.*

   Protective device: *Enter type and rating. For example, BS EN 60898 CB type B, 20 A.*

   Comments: *Note any defects/faults/omissions in other parts of the installation seen while conducting the minor works.*

3. Tests:

   Earth continuity: *Measured and then tick in box if OK.*

   Insulation resistance: *Standard tests and results.*

   EFLI (Z_s): *Standard tests and results.*

   Polarity: *Standard tests and then tick in box if OK.*

   RCD: *Standard tests, record operating current and time.*

4. Declaration

   Name, address, signature, etc.

CONTENTS OF A TYPICAL SCHEDULE OF TEST RESULTS

1. Contractor:	*Full name of tester.*
2. Date:	*Date of test.*
3. Signature:	*Signature of tester.*
4. Vulnerable equipment:	*Dimmers, electronic timers, CH controllers, etc. (i.e. anything electronic).*
5. Address:	*Full, or if in a large installation, the location of a particular DB.*
6. Z_e at origin:	*Measured value.*
7. PFC:	*Record the highest value that is PEFC or PSCC (should be the same for TN-C-S).*

8. Confirmation of supply polarity and phase sequence:	*Tick box.*
9. Instruments:	*Record serial numbers of each instrument, or one number for a composite instrument.*
10. Circuit Description:	*Details of circuit, e.g. first floor lights.*
11. kVA rating of protection:	*Taken from device.*
12. Type and rating:	*For example, BS EN 60898 CB type B, 32 A, or BS 88 40 A, etc.*
13. Reference method:	*Enter 'A', 'B', 'C', '100', etc.*
14. Wiring conductors:	*Size of live and cpc, e.g. 2.5 mm²/1.5 mm².*
15. Test results:	*Enter all measured results, not corrected values.*

SCHEDULE OF INSPECTIONS (AS PER BS 7671)

For EICs do not leave boxes uncompleted: N/A in a box if it is not relevant; ✓ in a box if it has been inspected and is OK. For EICRs, the same as the EIC except when defects, omissions, etc. are found, then enter one of the appropriate codes C1, C2, C3, or F1.

2394 Sample Paper

SECTION A

1. a. For the initial verification of a new installation, list six items of information regarding the incoming supply and three items regarding the installation that should be made available to the Inspector. (9 marks)

   b. State the three reasons for the issue of an Electrical Installation Certificate. (3 marks)

   c. State the documentation that must accompany an Electrical Installation Certificate. (2 marks)

   d. State the Statutory document that relates to Inspection and Testing. (1 mark)

2. a. State four methods of providing Basic Protection only.
      (4 marks)

   b. State two methods of providing both Basic and Fault Protection. (2 marks)

   c. Give an example of Electrical Separation in a domestic installation. (1 mark)

   d. State the IP codes applicable to an accessible horizontal top surface of an enclosure. (2 marks)

   e. State the conditions that would allow 'routing of cables in pre-scribed zones' to be used as the only method of protection where cables are buried less than 50 mm from the surface of a wall. (3 marks)

   f. State the three tests that require the use of a low-resistance ohmmeter. (3 marks)

3. a. Explain, with regard to ring final circuit continuity

      i. why, when Ls and Ns have been cross connected, the readings at each socket may vary considerably around the ring; (2 marks)

      ii. the significance of the highest reading when L and cpc are tested at each socket. (2 marks)

   b. List the first five tests, as per BS 7671:2008, to be conducted
      on a new light industrial unit where all circuits are radials. The
      supply earthing system is TN-C-S.   (5 marks)
   c. State two methods of providing additional protection. (2 marks)
   d. Explain in detail why an RCD test should not be conducted
      before an earth fault loop impedance test.   (4 marks)

SECTION B

All questions in this section relate to the following scenario:

A village cricket club has had a new clubhouse built comprising changing
rooms, a main office, a lounge and bar area, a players' waiting and viewing
area, and male and female toilets and shower facilities.

The building is a traditional brick and timber construction with
internal block walls. All cables are 70°C thermoplastic flat sheathed
6242/3Y, 'chased in' to full plaster depth and accessories are flush
mounted. All circuits are RCBO protected to BS EN 61009-1 type B. The
main protective bonding to gas and water services is 10 mm^2 copper. The
maximum length of this bonding conductor is 10 m.

The supply is 230 V 50 Hz single-phase and the earthing system is
TN-C-S with a measured Z_e value of 0.23 Ω and a PFC of 1 kA. The distribution fuse board is 12-way metal-clad.

Table A1.1 gives circuit details, Table A1.2 indicates conductor resistances.

All tests are carried out at 20°C.

Table A1.1						
Circuit No.	Device 30 mA	Description	Conductor Size mm^2		Length m End to End	Max Z_s
			Live	cpc		
1	32 A	Ring. Changing room & Office	2.5	1.5	20	1.37
2	32 A	Ring. Player area & Lounge	2.5	1.5	22	1.37
3	32 A	Ring. Kitchen	2.5	1.5	15	1.37

Circuit No.	Device 30 mA	Description	Conductor Size mm²		Length m End to End	Max Z_s
			Live	cpc		
4	40 A	Cooker. Kitchen	6.0	2.5	8	1.09
5	20 A	Hand dryer Male W/C	2.5	1.5	10	2.19
6	20 A	Hand dryer Female W/C	2.5	1.5	7	2.19
7	6 A	Lts. Changing room/ Office/Lounge	1.0	1.0	12	7.28
8	6 A	Lts. Player area & Kitchen	1.0	1.0	14	7.28
9	6 A	Lts. Male & Female W/C areas	1.0	1.0	16	7.28
10		Spare				
11		Spare				
12		Spare				

Table A1.2

Conductor Size mm²	Resistance in mΩ/m at 20°C
1.0	18.1
1.5	12.1
2.5	7.41
6.0	3.08
10.0	1.83

4. a. State the method used for protection against shock for this installation. (2 marks)

b. Explain why the lighting, cooker and hand dryer circuits have been RCD protected. (4 marks)

c. Explain the considerations to be made regarding the entry of the main tails and earthing conductor into the distribution board. (5 marks)

d. Determine the expected resistance R_2 of the main protective bonding conductor. (4 marks)

5. a. The kitchen ring final circuit has been securely isolated. Describe in detail how a ring final circuit continuity test would be carried out. (10 marks)

   b. What would be:

   i. the expected L to N resistance at each socket after correct L–N cross-connection? (2 marks)

   ii. the expected L to cpc resistance at each socket after correct L–cpc cross connection? (3 marks)

6. a. Draw a fully labelled diagram of the earth fault loop path for circuit 5. (8 marks)

   b. Explain why the PEFC and the PSCC for this installation would be the same value. (2 marks)

   c. Determine the expected Z_s values for circuits 1, 6 and 9 and confirm if they are acceptable. (5 marks)

2395 Sample Paper

SECTION A

1. a. State the four responsibilities of an inspector whilst carrying out a periodic inspection and test of an installation. (4)

   b. State the action to be taken by an inspector prior to carrying out a periodic inspection and test of a complex installation, where no diagrams, drawings, charts, etc. are available. (5)

   c. List the three documents that must be completed for the periodic inspection and test of an installation. (3)

   d. State one Statutory and two non-statutory publications that would apply to the periodic inspection and test of an installation. (3)

2. a. State the human senses that would be used when inspecting a metal conduit system in a 30-year-old installation. (2)

   b. List three considerations that would determine the frequency of periodic inspection and testing of an installation. (3)

   c. List three requirements of H&SE GS 38 regarding instrument test leads. (3)

   d. Describe the procedure for the safe isolation of supplies. (7)

3. a. State the action to be taken, giving reasons, regarding the earthing conductor during a test of external earth fault loop impedance. (2)

   b. Determine, from the following test results, which circuits do not comply with the requirements of BS 7671 (show all calculations):

Circuit No.	Measured Z_s Ω	Maximum Tabulated Z_s Ω
1	1.37	1.75
2	2.15	2.73
3	0.68	0.87
4	6.0	7.28

| 5 | 1.72 | 2.18 |
| 6 | 0.5 | 0.54 |

(3)

c. Describe, indicating all test values, the procedure for carrying out an instrument test on a 20 mA RCD used for additional protection. (10)

SECTION B

A small 30-year-old dry-powder coating company is situated adjacent to an 11 kV/400 V/230 V Distribution Network Operator's substation. The earthing system is T.N-C-S with an external earth fault loop impedance (Z_e) of 0.015 and a prospective fault current (PFC) of 15 kA. The Supply is three-phase four wire with 100 A BS 1361 fuses and the service tails are 35 mm² with a 25 mm² earthing conductor.

Gas and water services are located some 30 m from the electricity intake and no main protective bonding conductors are present.

At the intake position, a TP&N distribution board, housing BS 88 fuses, is used to supply all final circuits in the factory area. An SP&N consumer unit housing BS 3036 fuses is located next to the distribution board and supplies final circuits in a small kitchen area.

The wiring system throughout is 70°C thermoplastic, copper single core conductors enclosed in metal trunking and conduit.

The process of dry-powder coating of metal parts involves the use of acid baths, water jets, a tunnel oven and electrostatically charged powder spraying.

New owners have requested a periodic inspection and test of the whole installation before they continue with the production processes.

All original documentation, diagrams, charts, etc. have been lost.

All testing is carried out at a temperature of 20°C.

Table A1.3 gives resistance values of copper conductors at 20°C.

Table A 1.3   Conductor resistance

Conductor csa mm²	mΩ/m at 20°C
1.5	12.1
2.5	7.41
4.0	4.61
6.0	3.08
10.0	1.83
16.0	1.15

4. a. State three considerations to be made by the inspector in order to determine the initial amount of sampling to be carried out. (3)

   b. Explain the circumstances that may require an increase in the sampling rate above the initial amount determined. (3)

   c. State three circumstances where cables would not be inspected and would, therefore, not be included in the extent and limitations section of the Electrical Installation Condition Report. (3)

   d. List, apart from age, three general areas of investigation to be considered by the inspector carrying out the inspection of this installation. (3)

   e. State three external influences that should be considered for this installation. (3)

5. a. Explain why the protective devices in the SP&N consumer unit do not comply with BS 7671:2008. Indicate the classification code to be recorded. (5)

   b. State the cross-sectional area of the main protective bonding conductors that should be installed, the classification code to be recorded due to their omission, and the expected test resistance if they were installed. (6)

   c. List, apart from socket outlets not exceeding 20 A, four situations that require additional protection by RCDs. (4)

6. a. Describe how a test for cpc continuity, hence obtaining a value for $(R_1 + R_2)$, would be carried out on a radial circuit supplying a single-phase spray-booth motor. This circuit has been correctly isolated at its origin. (9)

b. Calculate the expected voltage drop in the motor circuit in (a) above, if the motor has a design current of 20 A and is supplied by 4 mm² live conductors with a measured line to neutral resistance of 0.15 Ω. (6)

2394 Sample Paper (Answers)

SECTION A

1. a. any six from: Earthing system
                    Number and type of live conductors
                    Voltage
                    Frequency
                    PFC
                    Z_e
                    Type and rating of supply protective device

   Any three from: Type and composition of wiring, type, number of points, etc.
                    Method of fault protection
                    Identification of protective devices, and isolators, switches, etc.
                    Details of circuits or equipment vulnerable to a typical test
                    Details of protective bonding conductors

   b. New installation; Addition; Alteration
   c. Schedule of test results; Schedule of inspections
   d. EAWR

2. a. Insulation; Barriers or enclosures; Obstacles; Placing out of reach
   b. SELV; PELV; Double insulation; Reinforced insulation (any two)
   c. Shaver point
   d. IPXXD or IP4X
   e. Installation must be under the supervision of a skilled or instructed person
   f. Continuity or protective conductors; Ring circuit continuity; Dead polarity

3. a. i.   Incorrect cross-connection
       ii. $R_1 + R_2$ for the ring
   b. Continuity of protective conductors
       1.   Insulation resistance

17th Edition IET Wiring Regulations: Inspection, Testing and Certification. 978-1-138-84886-3.

    ii. Polarity

    iii. Earth fault loop impedance

    iv. Operation of RCDs

  c. RCD; Supplementary equipotential bonding

  d. If the RCD test was done first and the earth loop path was too high, a fault would have been placed on the circuit and the RCD may not operate, placing everyone at risk

SECTION B

4. a. Automatic disconnection of supply

  b. These circuits only need RCD protection because the cables supplying them are less than 50 mm from the surface of the walls

  c. The tails and earthing conductor should enter the DB through the same hole to avoid electromagnetic effects. The entry hole should be to IPXXD or IP4X if top entry, IPXXB or IP2X elsewhere and cables protected against mechanical damage.

  d. 10 m of 10 mm^2 bonding conductor $= \dfrac{10 \times 1.83}{1000} = 0.018 \, \Omega$

5. a. • Select a low-resistance ohmmeter

    • Check for damage and battery condition and current calibration

    • Null or zero test leads (GS 38 leads)

    • Identify legs of ring

    • Measure individual loops, r_1, r_n and r_2

    • Cross-connect opposite Ls and Ns

    • Measure between L and N at each socket (should be the same value)

    • Repeat with Ls and cpcs crossed (highest reading is $R_1 + R_2$)

    • Record $(R_1 + R_2)$

    • Reinstate system

  b. i. r_1 and r_n each $= \dfrac{15 \times 7.41}{1000} = 0.11 \, \Omega$ so reading at each socket

    $= \dfrac{0.11}{2} = 0.055 \, \Omega$

ii. $r_2 = \dfrac{15 \times 12.1 \times 1.2}{1000} = 0.18\ \Omega$ so reading at each socket

$\qquad = \dfrac{0.11 + 0.18}{4} = 0.073\ \Omega$

6. a.

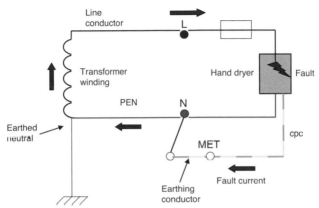

4 marks for diagram, 4 marks for labels (half each)
NO MARKS IF DIAGRAM INCORRECT

b. The earth fault current and the short-circuit current take the same path, i.e the PEN conductor

c. CCT 1 $R_1 + R_2 = \dfrac{(7.41 + 12.1) \times 20}{4 \times 1000} = 0.1\ \Omega$.

$Z_s = 0.23 + 0.1 = 0.33\ \Omega$ ok, $0.8 \times$ max is $1.09\ \Omega$

CCT 6 $R_1 + R_2 = \dfrac{(7.41 + 12.1) \times 7}{1000} = 0.13\ \Omega$.

$Z_s = 0.23 + 0.13 = 0.36\ \Omega$ ok, $0.8 \times$ max is $1.75\ \Omega$

CCT 9 $R_1 + R_2 = \dfrac{(18.1 + 18.1) \times 16}{1000} = 0.58\ \Omega$.

$Z_s = 0.23 + 0.58 = 0.81\ \Omega$ ok, $0.8 \times$ max is $5.83\ \Omega$

2395 Sample Paper (Answers)

SECTION A

1. a.   • Ensure their own safety and that of others, etc. (1)
      - Compare results with relevant criteria (1)
      - Confirm compliance with BS 7671 (1)
      - Take a view of the condition of the installation (1)   (4 marks)
   b. Exploratory survey to establish if safe to carry out inspection and testing without diagrams/drawings, etc. If not, then the client can be required to produce them.   (5 marks)
   c. EICR, schedules of inspections and schedules of test results
          (3 marks)
   d. EAWR, BS 7671, GN3, On-site guide, etc.   (3 marks)

2. a. Touch and sight   (2 marks)
   b. Type of installation and equipment; use of installation; external influences; quality and frequency of maintenance (any three)
          (3 marks)
   c. No more than 4 mm probe tips; finger guards; coloured; suitably insulated; fused; suitable length (any four)   (3 marks)
   d.   • Seek permission to isolate (1)
      - Isolate and lock off (1)
      - Retain key (1)
      - Select an approved voltage indicator to GS38 (1)
      - Prove indicator on known supply with same voltage as system under test (1)
      - Check system is de-energized (1)
      - Reprove voltage indicator (1)   (7 marks)

3. a. Disconnect to avoid parallel paths   (2 marks)
   b.

Circuit No.	Measured Z_s Ω	Corrected Maximum Z_s Ω
1	1.37	1.75 × 0.8 = 1.4
2	2.15	2.73 × 0.8 = 2.18
3	0.68	0.87 × 0.8 = 0.7

4	6.0	7.28 × 0.8 = 5.82
5	1.72	2.18 × 0.8 = 1.74
6	0.5	0.54 × 0.8 = 0.43

Circuits 4 and 6 do not comply   (3 marks)

c. • Ensure earthing is satisfactory (1)
   • Select RCD tester (1)
   • Check leads are to GS38 (1)
   • Connect leads to L, N and E on load side of RCD (1)
   • Test at 20 mA on both 0° and 180° (1)
   • Check RCD trips within 300 ms (1)
   • Record longest time (1)
   • Repeat test at 100 mA on both 0° and 180° (1)
   • Check RCD trips within 40 ms (1)
   • Record longest time (1)   (10 marks)

SECTION B

4. a. Any three from the list on 'Setting inspection and testing samples' GN3   (3 marks)
   b. If inspection or testing of the original sample reveals poor results   (3 marks)
   c. Cables in conduit or trunking; under floors; in roof spaces; within the building fabric or underground (any three) (3 marks)
   d. Any other three from the SADCOWES list   (3 marks)
   e. Water; dust; corrosion; heat (Any three)   (3 marks)

5. a. BS 3036 fuses have a very low breaking capacity (4 kA max) and the PSCC at the origin, where these fuses are located, is 15 kA; hence a short-circuit on any of the circuits fed by these fuses is likely to cause serious damage/arcing/fire in the SP&N consumer unit   (5 marks)
   b. 10 mm²; C2; $\dfrac{1.83 \times 30}{1000} = 0.055\ \Omega$ (6 marks)
   c. Circuits supplying mobile equipment not exceeding 32 A, outdoors; Cables in walls less than 50 mm; All LV circuits in

bathrooms; Circuits in saunas; Swimming pools; Agricultural locations, etc.   (4 marks)

6. a.   ▪ Isolate/disconnect at motor end of circuit (1)
       ▪ Select a low-resistance ohmmeter (1)
       ▪ Check for instrument damage (1)
       ▪ Check battery condition (1)
       ▪ Check leads are to GS 38 (1)
       ▪ Link line to cpc at DB or motor (1)
       ▪ Test between line and cpc at other end (1)
       ▪ Record result
       ▪ Reinstate system (1)   (9 marks)

   b.   Circuit resistance = $0.15 \times 1.2 = 0.18 \ \Omega$
      Motor current = 20 A
      Volt drop = $20 \times 0.15 \times 1.2 = 3.6$ volts   (6 marks)

2394/5 Sample MCQ Paper and Answers

1. In order to comply with BS 7671, which one of the following should an inspector be?
   a. An approved contractor
   b. A competent person
   c. A qualified electrician
   d. An instructed person

2. Which one of the following fulfils the requirement for safe isolation of a circuit?
   a. Switch off and inform all personnel
   b. Switch off and display notices
   c. Lock off and display notices
   d. Lock off and give key to client

3. Which one of the following is the correct procedure for the use of a voltage indicator when carrying out safe isolation?
   a. Check the indicator via its own test button
   b. Check the indicator on the isolated circuit
   c. Check the indicator on a proving unit
   d. Check the indicator is connected to earth

4. Which one of the following is the correct action for an inspector after completing the safe isolation of a circuit?
   a. Keep the key in a tool-box
   b. Hand key to the client
   c. Give duplicate key to client
   d. Keep key on person

5. The supply to an office complex is to be isolated in order to carry out an inspection and test. Which one of the following is the inspector's first step in the isolation process?
   a. Switch off and prove dead
   b. Seek permission to isolate

    c. Carry out a continuity test

    d. Inform the energy supplier

6. A property which includes a solar PV system is to be inspected and tested. Which one of the following actions must be taken to ensure safe isolation?

    a. Isolate both main and PV supplies

    b. Isolate PV supply only

    c. Isolate main supply only

    d. Link out the PV metering

7. Which one of the following can only be conducted with the supply isolated?

    a. RCD test

    b. Phase rotation at a motor

    c. Ring final circuit continuity

    d. Earth fault loop impedance

8. Which one of the following is the status of BS 7671?

    a. Statutory

    b. Non-statutory

    c. Advisory

    d. Mandatory

9. Which one of the following would be the most suitable as a reference for an inspector conducting an inspection and test?

    a. BS 7671

    b. GS38

    c. EAWR

    d. GN3

10. Which one of the following documents requires the details of the method of fault protection?

    a. Electrical installation certificate

    b. Electrical installation condition report

    c. Minor works certificate

    d. Schedule of test results

11. GS38 requires that low voltage test leads should be:
    a. fused, have finger guards and no more than 4 mm exposed tips
    b. fused, have finger guards and more than 4 mm exposed tips
    c. non-fused, have finger guards and no more than 4 mm exposed tips
    d. non-fused, have finger guards and no more than 2 mm exposed tips

12. Which one of the following is the instrument to be used for an insulation resistance test?
    a. A resistance tester
    b. A high-reading ohmmeter
    c. an ohmmeter
    d. an insulation resistance tester

13. Which one of the following units is displayed on an RCD tester?
    a. ms
    b. Ms
    c. mA
    d. MΩ

14. The technical specification for a low-resistance ohmmeter requires the instrument to deliver a no-load voltage and a short-circuit current:
    a. between 4 V–24 V a.c.; not exceeding 200 mA
    b. between 4 V–24 V d.c.; not less than 100 mA
    c. between 4 V–24 V a.c. or d.c.; not less than 200 mA
    d. between 4 V–24 V a.c. or d.c.; not exceeding 200 mA

15. The following tests are to be conducted on an installation:
    i. Earth fault loop impedance; ii. RCD operation; iii. Continuity of protective conductors; iv. insulation resistance.
    Which one of the following is the correct sequence for these tests?
    a. 1, 2, 3, 4
    b. 3, 2, 1, 4
    c. 4, 3, 2, 1
    d. 3, 4, 1, 2

16. Which one of the following tests automatically conducts a dead circuit polarity test?
    a. Insulation resistance
    b. Ring final circuit continuity
    c. Protective bonding continuity
    d. RCD operation

17. Which one of the following tests requires the instrument leads to be tested 'open' and 'shorted' before testing commences?
    a. Insulation resistance
    b. Conductor continuity
    c. RCD operation
    d. Polarity

18. Which one of the following gives details of the correct use of a voltage indicator?
    a. GN3
    b. BS 7671
    c. GS38
    d. EAWR

19. In order to significantly reduce the risk of electric shock under 'fault-free' conditions, it is important that:
    a. disconnection times are met
    b. basic protection is provided
    c. the earthing conductor is ≥ 6.0 mm^2
    d. a 'non-standard colours' notice is displayed

20. A test for continuity of a main protective bonding conductor requires its disconnection at one end. This is to eliminate:
    a. the risk of shock during the test
    b. the test lead resistance
    c. excessive volt drop during the test
    d. the introduction of parallel paths

21. Which one of the following would be the effect on the resistance of a cable if both its csa and length were doubled?
    a. Halved
    b. Stayed the same
    c. Doubled
    d. Quadrupled

22. A length of 2.5 mm² flat sheathed cable has end-to-end live conductor resistances of 0.37 Ω. Which one of the following would be the approximate resistance of the associated 1.5 mm² cpc?
    a. 1.67 Ω
    b. 1.48 Ω
    c. 0.62 Ω
    d. 0.19 Ω

23. Which one of the following would be the overall value of insulation resistance of an installation if individual circuit values were 10 MΩ; 15 MΩ; 25 MΩ and 45 MΩ?
    a. 95 MΩ
    b. 23.75 MΩ
    c. 4.37 MΩ
    d. 0.23 MΩ

24. Which one of the following would **not** be recorded on a schedule of test results for a ring final circuit continuity test?
    a. r_1
    b. r_n
    c. L to N at each socket
    d. L to cpc at each socket

25. Which one of the following would be the effect on the insulation resistance of a cable if both its csa and length were doubled?
    a. Halved
    b. Stayed the same
    c. Doubled
    d. Quadrupled

26. Which one of the following earthing systems includes a PEN conductor?
   a. TT
   b. TN-S
   c. IT
   d. TN-C-S

27. When conducting an earth electrode resistance test on a generator electrode, an acceptable value should be the average of three readings and an instrument accuracy of:
   a. 100%
   b. >5%
   c. >2%
   d. ≤2%

28. An external earth fault loop impedance test must be carried out with the earthing conductor:
   a. disconnected and the installation energized
   b. connected and the installation energized
   c. disconnected and the installation isolated
   d. connected and the installation isolated

29. When conducting an earth fault loop impedance test on a radial circuit the earthing conductor of the installation must be:
   a. disconnected and all protective bonding conductors in place
   b. disconnected and all protective bonding conductors disconnected
   c. connected and all protective bonding conductors in place
   d. connected and all protective bonding conductors disconnected

30. Which one of the following would be the maximum value of Z_s permitted to achieve the required disconnection time, for a maximum measured value of earth fault loop impedance of 1.15 Ω?
   a. 2.3 Ω
   b. 1.92 Ω
   c. 1.44 Ω
   d. 0.92 Ω

31. Which one of the following needs to be compensated for when comparing measured values of Z_s with maximum tabulated values?
    a. Cable length
    b. Cable csa
    c. Change in temperature
    d. Type of protective device

32. The value of Z_s to be recorded for an earth fault loop impedance test on a ring final circuit is the value measured at a socket outlet:
    a. nearest the distribution board
    b. that gives the highest reading
    c. nearest the mid point of the ring
    d. that gives the lowest reading

33. Which one of the following is the requirement for RCDs used for additional protection?
    a. >30 mA and tripping in 40 ms at $5 \times I_{\Delta n}$
    b. ≤30 mA and tripping in 40 ms at $5 \times I_{\Delta n}$
    c. <30 mA and tripping in 40 ms at $5 \times I_{\Delta n}$
    d. ≥30 mA and tripping in 300 ms at $5 \times I_{\Delta n}$

34. Which one of the following is the maximum test current required for a 300 mA RCD installed for protection against fire?
    a. 1500 mA
    b. 300 mA
    c. 100 mA
    d. 30 mA

35. Which one of the following is the rating of an RCD installed for additional protection if the maximum test current is 100 mA?
    a. 100 mA
    b. 30 mA
    c. 20 mA
    d. 10 mA

36. When conducting a prospective fault current test at the origin of an installation, the earthing conductor must be:
    a. disconnected
    b. linked to the neutral
    c. connected
    d. excluded from the test

37. A prospective fault current test at the origin of a three-phase installation gave the following results: PEFC – 4.7 kA; PSCC – 5.5 kA. Which one of the following would be recorded on a schedule of test results?
    a. 11 kA
    b. 10.2 kA
    c. 5.5 kA
    d. 4.7 kA

38. It is recommended that a functional test on an RCD via its integral test button, be carried out:
    a. annually
    b. half-yearly
    c. every three months
    d. every four months

39. Disconnection times exceeding the permitted maximum values are a direct result of:
    a. insulation resistance values approaching their minimum
    b. poorly connected main protective bonding conductors
    c. earth fault loop impedance values exceeding BS 7671 requirements
    d. earth fault loop impedance values unacceptably low

40. A leakage current of 11 mA between L and E occurring in a 230 V circuit would be the result of a L–E insulation resistance value of:
    a. 2.53 MΩ
    b. 0.02 MΩ
    c. 0.5 MΩ
    d. 1.0 MΩ

ANSWERS

1 **b**	11 **a**	21 **b**	31 **c**
2 **c**	12 **d**	22 **c**	32 **b**
3 **c**	13 **a**	23 **c**	33 **b**
4 **d**	14 **c**	24 **c**	34 **b**
5 **b**	15 **d**	25 **a**	35 **c**
6 **a**	16 **b**	26 **d**	36 **c**
7 **c**	17 **a**	27 **d**	37 **a**
8 **b**	18 **c**	28 **c**	38 **c**
9 **d**	19 **b**	29 **c**	39 **c**
10 **c**	20 **d**	30 **c**	40 **b**

Index

Page numbers in *italics* denote an illustration, **bold** indicates a table